PRAISE FOR **AGROECOLOG**

MW01493609

"In *Agroecology and Regenerative Ag*
industrialized agricultural product
environment, the economy, and human health. We cannot afford to continue down
this path. Shiva shows us how imperative it is for humankind to embrace Regenerative
Agriculture. A must-read for all who are concerned about our future."

— Gabe Brown, Farmer, Rancher, Educator, and author, *Dirt to Soil: One Family's
Journey into Regenerative Agriculture*

"Once again, a relentless and passionate Shiva brings light to the latest in colonialism and
nefarious Wiindigoo capitalism. In their rabid consumption of life forms, the Wiindigoos
destroy the world that feeds them. We find ourselves at a crossroads, between two
paths—one green and the other scorched. It is known as the Seventh Fire. And, at this
moment, Vandana reminds us that we still have seeds and soil. In those life forms is great
hope and promise."

— Winona LaDuke, Economist, Environmentalist, and Executive Director, Honor
the Earth

"With all her usual power and clarity, Shiva describes how we can transition to a food
system that regenerates ecosystems, builds biodiversity, and provides healthy nutrition
for all—starting with soil, seeds, farms, and food. This book brings together years of
research and experience from the world's leading ecofeminist scholar. Don't miss it."

— Jason Hickel, Economic Anthropologist and author, *Less is More: How Degrowth
Will Save the World*

"*Agroecology and Regenerative Agriculture* integrates cutting edge science with indigenous
wisdom into a clarion call for the future of agriculture. The methods, principles, and case
studies in this book can't be ignored. While others debate endlessly, Vandana Shiva and
her team do the work and prove this science—THIS is how we end world hunger and
mitigate climate catastrophe."

— Erik Ohlsen, Executive Director, Permaculture Skills Center

"Vandana Shiva is the defining voice of the organic regenerative agriculture movement.
Here she lays out how the misnamed 'Green Revolution' advanced an industrial
agriculture based on poisons and patents that are killing people and the planet to grow
the profits of transnational corporations. An essential read for anyone committed to
advancing the emergence of an Ecological Civilization dedicated to the well-being of
people and Earth."

— David Korten, author, *When Corporations Rule the World* and *The Great Turning:
From Empire to Earth Community*

"Based upon decades of ethnoecological research and farming experience, this comprehensive and path-breaking book brings the Traditional Ecological Knowledge of farmers to the forefront. It underscores the importance of dozens of actions that farmers do or can do, which in turn work with the renewal capacity of the earth while securing the health of the global commons. This book is relevant, now more than ever, given the globalization agenda, the loss of biodiversity, food sovereignty, pollution, and climate change around the world."

— M. Kat Anderson, author of *Tending the Wild: Native American Knowledge and the Management of California's Natural Resources*

"An enlightened synthesis that weaves together decades of experience and studies from around the world to challenge the foundational myths of modern agriculture. With a scientist's eye for insight and an activist's passion for change, Shiva lays out an urgent call for ecologically-based regenerative farming rooted in restoring diversity and organic matter."

— David R. Montgomery, author of *Dirt: The Erosion of Civilizations* and *Growing a Revolution: Bringing Our Soil Back to Life*

"Vandana Shiva skillfully pulls together threads that for too long have been treated as separate epidemics; from human health and well-being, food production, to environmental and global climate change. Bringing together over 30 years of real-world practice and research at Navdanya, Shiva celebrates the important role of farmers as key figures in regeneration and reversing the far-reaching catastrophic consequences from the failed promises of the 'Green Revolution.'"

— Nicole Masters, Agroecologist and author of *For the Love of Soil: Strategies to Regenerate our Food and Production Systems*

"Highlighting many examples of agrarian distress within land and people, this book is a much-needed clarion call for why biodiverse organic farming makes sense at a human and planetary level. It is a well-researched volume and a skillful balance of breadth and depth, scholarship and hands-on guidance borne of lived-experience; Vandana Shiva's depth of understanding sings through every page."

— Jane Raddiford, PhD, author of *Learning to Lead Together: An Ecological and Community Approach*

"*Agroecology and Regenerative Agriculture* is a magnificent manifesto for the future of good farming; which can sequester carbon dioxide, feed everyone with nutritious food, preserve the integrity and dignity of farmers, and restore biodiversity. This book is urgently needed. Everyone should read it."

— Satish Kumar, Founder of Schumacher College and Editor Emeritus, *Resurgence & Ecologist*

Agroecology & Regenerative Agriculture

Agroecology & Regenerative Agriculture

Sustainable Solutions for Hunger, Poverty, and Climate Change

✦ ✦ ✦ ✦ ✦

VANDANA SHIVA

FOREWORD BY
HANS R. HERREN

SYNERGETIC PRESS

SANTA FE ✦ LONDON

Synergetic Press | 1 Bluebird Court, Santa Fe, NM 87508
& 24 Old Gloucester St. London, WC1N 3AL England

Library of Congress Cataloging-in-Publication Data
Names: Shiva, Vandana, author.
Title: Agroecology and regenerative agriculture : sustainable solutions for hunger, poverty, and climate change / Vandana Shiva.
Other titles: Sustainable solutions for hunger, poverty, and climate change

Description: Santa Fe, NM : Synergetic Press, [2022] | Summary: "This book is an interdisciplinary synthesis of research and practice carried out over decades by leaders of the agroecology and regenerative organic agriculture movement. It provides detailed analysis of the multiple crises we face due to chemical and industrial agriculture, including land degradation, water depletion, biodiversity erosion, climate change, agrarian crises, and health crises. The book lays out biodiversity based organic farming and agroecology as the road map for the future of agriculture and sustainable food systems (both locally and globally). With detailed scientific evidence, Agroecology & Regenerative Agriculture shows how ecological agriculture based on working with nature rather than abasing ecological laws can regenerate the planet, the rural economy, and our health"-- Provided by publisher.
Identifiers: LCCN 2021054025 | ISBN 9780907791935 (paperback) | ISBN 9780907791942 (ebook)
Subjects: LCSH: Biodiversity. | Agricultural ecology. | Sustainable agriculture. | Organic farming.
Classification: LCC QH541.15.B56 S55 2022 | DDC 333.95--dc23/eng/20211123
LC record available at https://lccn.loc.gov/2021054025

ISBN 9780907791935 (paperback)
ISBN 9780907791942 (ebook)

Cover Design by Amanda Müller
Cover Illustration by Leticia Pascual
Book design by Brad Greene
Managing Editor: Amanda Müller
Project Editor: Sage Wylder
Printed in the USA

TABLE OF CONTENTS

SECTION 1

SECTION 2

SECTION 3

SECTION 6

SECTION 7

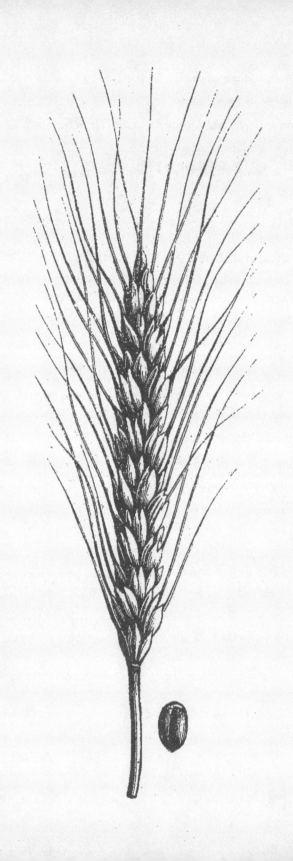

Hans R. Herren

President, Millennium Institute, Founder & Chair, Biovision Foundation, and Co-chair, IAASTD,

♦ ♦ ♦

The timing for the publication of this insightful book could not be better, given the culmination of several crises affecting the food system, not the least the COVID 19 outbreak that quickly developed into a pandemic with the dire consequences we now know, and is far from over. The alarm had already been ringing for several decades, indeed many calls for a closer look at the direction of the food system development have been voiced, as far back as Silent Spring and Limits to Growth over 50 years ago.

The calls for radical changes in the way we grow our food, away from the synthetic chemicals to control pests and fertilize the crops, have been called ideologies and dreams that would be unable to feed a growing population. As we have just experienced, the globalized food system, depending on global trade of a few commodities, is indeed very vulnerable and prone to fail when supply chains are broken by events such as a pandemic, or other factors such as rising energy costs and major weather events among others.

The state of our food system is at best insufficient. In reality, it has been designed by greed and the knowledge that controlling food is controlling people. This led, on the one hand, to keep food prices as low as possible, in particular the staples such as wheat (bread), maize, sugar, palm oil, while the concentration on only a few commodities allowed to streamline the supply chains, and so keep prices low. These low prices also kept the farmers in a dependent relationship with the suppliers of inputs, and of the buyers of their products, therefore squeezing them between a rock and a hard place. It's no wonder that as the food system developed, the number of farmers diminished, often through bankruptcies and the inability to keep up with debts arising from low product, high input prices, and the ever-increasing price of land. The resulting picture is "get big or get out" with mega farms, mega CAFOs, mega machines, mega environmental pollution, mega biodiversity loss, mega climate change contributions on the one hand, and on the other mini income, mini nutrition and health. Clearly such a system cannot work in the medium and long term, neither from economic, social, or environmental considerations. One dramatic example not to repeat is the pal palm oil industry in Indonesia, but about to repeat itself in the Democratic Republic of Congo, where the enormous peatland areas are targeted for exploitation. How much more destruction of pristine and unreplaceable nature, biodiversity, will be sacrificed to support an ever-widening income gap, as clearly the benefits will not accrue to the local people?

As described in the IPES-Food report "From Uniformity to Diversity," the key factors that block a transformation of the food system towards the agroecological principle (as defined in the latest CFS HLPE report on Agroecological and other Innovative Approaches that enhance food security and nutrition) are the power concentration at the center of the globalized food system, and the connected short term thinking—expectation of cheap food, path dependency, export orientation, feed the world narrative, measure of success, and compartmentalized thinking. Transforming the food system means to tackle all of these elements together. There is ample evidence that alternative practices exist all along the food value chain, from production to consumption, and the return loop to the soil for waste as the readers will learn from this book. From agroecological practices, including organic regenerative and permaculture, which do not use chemical input and genetically modified seeds, there are proven so(i)lutions for farmers to choose from to produce healthy food in a healthy and diverse environment. These practices can, and will when implemented across the world, while also being mindful of energy use, be neutral at worst and positive most likely in terms of climate change promoting emissions. The cheap food drive, which is most pronounced in the meat sector, is also the main contributor to CC and we will not be able to stay within the set target of 1.5 degrees Celsius unless changes in our diets are also undertaken.

The planetary boundaries, and the by now well documented overshoots, indicate that the food system is a major factor in the areas of biosphere integrity (functional and genetic diversity), nitrogen and phosphorus pollution, land-system change (degradation), and freshwater use (pollution). Will the decision makers pay attention? Likely only if the present crises persist and get worse. The great expectation that techno fixes will come to our rescue will not happen. As we have experienced it over the last half century, one techno fix calls for the next one, that's the essence of techno fixes, as they tackle the symptoms of the problem, instead of dealing with the root causes. In the case of the food system, please refer to the IPES-Food report mentioned above. Best examples of techno fixes are pesticides, herbicides, and fungicides, which bring farmers into a treadmill, from which only a transition to agroecological and organic regenerative practices can free them. The linked genetic engineering, inclusive of CRSPRcas is in the same vein and has not solved a single problem permanently, as for example biological control programs that use nature's gifts to solve problems.

That seeds, just as soil are to be returned to the ownership of the farmers, and especially women farmers, is one big tenet of the food system transformation. After all, farmers around the world have been selecting and saving the best ecologically and culturally adapted seeds, and provided care for the soil in which they grow their crops. Farmers need to be in charge of their farms and farm inputs, they need to have control over the prices to assure that the most relevant profession in the world gets out of being also the poorest and suffering from malnutrition. The effort to introduce fair prices is helping some farmers already, however, more can be done by reassigning billions in subsidies to farmers that are transforming their farms to agroecological practices, as well as to operators higher in the value chain for producing food that is nutritious and free of chemical residues. To get the proper pull from the consumers end, true cost accounting is an idea for which time has come. By accounting for both the positive and negative externalities in the final product price, today's cheap food would become more expensive than sustainably produced food. Farmers would also have an additional incentive to transition to agroecological practices.

With all the knowledge and science available to make the transformation of the food system a reality, what levers could be best applied? As detailed in this well-done book—the issues, backed up with solid data and references that affect the food system are elaborated on in great details, as are the crises that have emerged over the past decades. Furthermore, and very much to the point, solutions are highlighted with examples from Navdanya's research and implementation activities, defeating the perennial arguments from agribusiness and other big ag vested interests that an agroecology and organic-regenerative agriculture food-based system cannot nourish the world. Noteworthy is the reference to the need to be inclusive as well as integrated at all three sustainable and equitable development levels, social, environmental,and economic.

SECTION 1 Introduction

1.1 *Multiple Crises in Agricultural Systems and the Urgent Need for a Paradigm Change*

Among the many dangers facing humanity exists a particularly menacing triple crisis that threatens our agricultural and food systems. The first part of this is the ecological crisis, which includes the following:

+ The disappearance of biodiversity and species
+ Climate change, climate instability, and climate extremes
+ Soil erosion, land degradation, and desertification
+ Water depletion and pollution
+ The spread of toxins throughout the food system

The second is the public health crisis of hunger, malnutrition, and the non-communicable chronic disease epidemic. The third crisis concerns farmers' livelihoods, indebtedness, and suicides because of high-cost inputs and displacement due to land degradation and desertification.

All three crises are interconnected, even though they are generally seen as separate. The most significant contribution to all three crises comes from a fossil fuel-intensive, chemical-intensive, and capital-intensive system of non-sustainable industrial agriculture which is degrading the environment, public health, and farmers' livelihoods.

The gravity of these crises is a clear indication that the old paradigm of agriculture is clearly broken. As the UN Report of the International Assessment of Agriculture, Science, Technology, and Development (IAASTD) has noted, business as usual is no longer an option. Neither the Green Revolution nor GMOs can guarantee food security.

The industrial agriculture model was introduced on the grounds that it would increase food security by increasing food production and farmers' income. While the production of a handful of agricultural commodities has increased, the biodiversity of crops—which is vital to nutrition and health—has gone down. Industrial food is nutritionally empty and full of toxins, aggravating malnutrition and creating public health dilemmas.

5

These negative impacts on the planet and society are built into the scientific paradigm and technologies of industrial agriculture, many of which are rooted in the mindset of war. This mindset maintains a militaristic view of humans at war with nature, farmers competing with each other, and countries engaged in trade wars. The chemicals used in industrial farming have their origins in warfare, specifically, the gas chambers of concentration camps. The introduction of agrochemicals in the West nearly a century ago (and in India in the 1960s with the Green Revolution) changed the paradigm of agriculture, making it the biggest contributor to the degradation of the Earth. It focused on the external inputs of chemicals and neglected the role and function of biodiversity in living seeds and soil, along with the hydrological and nutrient cycles which maintain Earth's climate systems. Instead of working with ecological processes that are embodied in agroecology—considering the health of the entire agroecosystem and its diverse species—agriculture was reduced to an external input system and adapted to toxic chemicals.

Industrial agriculture has been imposed under the illusion that a paradigm based on war against the Earth is the only "science" available and that the use of war chemicals in farming is the best means to provide food security to humanity.

In contrast, a scientific and ecologically robust paradigm of agriculture is emerging in the form of biodiversity, agroecology, and regenerative organic farming, which addresses the triple crisis. Instead of degrading the soil, health, and rural livelihoods, it rejuvenates and regenerates them. Instead of using toxic chemical inputs which cause harm to the environment and public health, it relies on a diversity of flora, fauna, and microorganisms, each with respective ecological functions.

Biodiversity, agroecology, and regenerative organic farming are the ecological practices to address poverty, hunger, and multiple harms to public health that have been caused by chemical and fossil fuel-intensive industrial agriculture. This is the paradigm shift needed to meet the Sustainable Development Goals as outlined by the UN, especially goals 1, 2, and 3.

GOAL 1: No Poverty: End poverty in all its forms everywhere

"Poverty is more than the lack of income and resources to ensure a sustainable livelihood. Its manifestations include hunger and mal-nutrition, limited access to education and other basic services, social discrimination and exclusion, as well as the lack of participation in decision-making. Economic growth must be inclusive to provide sustainable jobs and promote equality."

GOAL 2: Zero Hunger: Achieve food security, improve nutrition, and promote sustainable agriculture

"It is time to rethink how we grow, share, and consume our food."

GOAL 3: Good Health and Well-Being: Ensure healthy lives and promote well-being for all at all ages

"Ensuring healthy lives and promoting the well-being for all at all ages is essential to sustainable development. Significant strides have been made in increasing life expectancy and reducing some of the common killers associated with child and maternal mortality."

In this emerging paradigm, the technologies of production are the ecological functions provided by biodiversity, also known as nature's laws of diversity and return. As Navdanya's practice and research over the past three decades have shown, by conserving and intensifying biodiversity in agro-ecosystems, we produce more food and nutrition; increase farmers' incomes, regenerate the soil, water, the biodiversity, and mitigate climate change by sequestering carbon from the atmosphere into the soil. That is why we call our system of farming *regenerative* organic agriculture.

This book synthesizes 31 years of Navdanya's practice and research on biodiverse organic farming. It shows that through biodiversity and agro-ecology, we can double the production of real food (based on nutrition per acre) and increase farmers' net incomes (based on wealth per acre and true-cost accounting). Our organization's research is complemented by Andre Leu's global experience as an organic farmer, the former Chair of the Board of IFOAM, and the current International Director of Regeneration International, an organization we co-founded with Hans Herren of IAASTD and Ronnie Cummins of the Organic Consumers Association.

1.2 *Degradation of the Environment, Public Health, and Rural Economies*

Industrial agriculture is responsible for the four interrelated environmental catastrophes facing the planet: the dramatic decline in biodiversity, the effects of climate change, land degradation, and the water crisis. Together, these crises contribute to what is known as the "Anthropocene Extinction," the sixth major extinction event on our planet. This decline in species—especially bees, birds, and frogs—has occurred primarily through the use of toxic agrochemicals that lower fertility and immunity, act as endocrine disruptors, cause birth defects, and other negative health effects.

The United Nations Millennium Ecosystem Assessment Synthesis Report is the most comprehensive study ever conducted into the state of the environment on the planet. This detailed report by many of the world's leading scientific experts showed that our current agricultural practices are clearly unsustainable and contributing to biodiversity loss. Over the past 50 years, humans have changed ecosystems more rapidly and extensively than in any comparable period of time throughout human history, largely to meet growing demands for food, freshwater, timber, fiber, and fuel. This has resulted in a substantial and largely irreversible loss of the diversity of life on Earth.

A 2001 study from the University of California stated that agriculture would be a major driver of global environmental change over the next 50 years, rivaling the effect of greenhouse gases in its impact. The lead author, David Tilman, found that the use of pesticides, chemical fertilizers, and habitat destruction have caused a major extinction event that is lowering the world's biodiversity and changing its ecology: "Neither society nor most scientists understand the importance of agriculture." Tillman states, "It's grossly misunderstood, barely on the radar screen, yet it is likely as important as climate change. We have to find wiser ways to farm."

The International Assessment of Agricultural Knowledge, Science, and Technology for Development (IAASTD) Synthesis Report was the largest review of our current global agricultural systems ever undertaken. This was a multi-stakeholder process that involved over 400 scientific authors, 61 countries, and a bureau co-sponsored by the United Nations Food and Agriculture Organization (FAO), the Global Environment Facility (GEF), United Nations Development Programme (UNDP), United Nations

Environment Program (UNEP), United Nations Educational, Scientific, and Cultural Organization (UNESCO), the World Bank, and the World Health Organization (WHO). Uncovered in the report were multiple environmental problems affecting the sustainability of global agricultural production:

A. Land degradation and nutrient depletion

- Land degradation occurs on about 2,000 million ha of land worldwide, affecting 38% of the world's cropland
- Land degradation has depleted nutrients in the soil, resulting in N, P, and K deficiencies covering 59%, 85%, and 90% of harvested area (respectively) in the year 2000
- This has resulted in a loss of 1,136 million tons per year in total global production
- 1.9 billion ha (and 2.6 billion people) today are affected by significant levels of land degradation

B. Salinity and acidification

- Salinization affects about 10% of the world's irrigated land

C. Loss of biodiversity (above and below ground) and associated agroecological functions

- Caused by repeated use of monoculture practices
- Excessive use of agrochemicals
- Agricultural expansion into fragile environments
- Excessive land clearance of natural vegetation adversely affecting productivity

D. Reduced water availability, quality, and access

- Fifty years ago, water withdrawal from rivers was one-third of what it is today
- Agriculture already consumes 75% of all global freshwater withdrawn worldwide

E. Increased pollution (air, water, land)

- Increasing pollution also contributes to water quality problems affecting rivers and streams

+ There have also been negative impacts of pesticide and fertilizer use on soil, air, and water resources throughout the world
+ Agriculture contributes about 60% of anthropogenic emissions of CH4 and about 50% of N20 emissions
+ Inappropriate fertilization has led to eutrophication and large dead zones in a number of coastal areas
+ Inappropriate use of pesticides has led to groundwater pollution, health problems, and loss of biodiversity

The IAASTD report concluded that our current agricultural production systems are unsustainable and need to change: "The way the world grows its food will have to change radically to better serve the poor and hungry if the world is to cope with growing population and climate change while avoiding social breakdown and environmental collapse." The GMO and industrial agriculture paradigms are not endorsed, and instead, the solutions suggested are to improve sustainability, work at a local level with lower inputs, and use ecological farming methods, including organic farming.

"Business as usual" is not an option if we want to achieve environmental sustainability. Industrial monoculture agriculture has pushed more than 75% of genetic plant diversity to extinction, with large masses of bees dying due to toxic pesticides. Albert Einstein had cautioned that when the last bee disappears, so will humans. As the FAO stated on the International Day for Biological Diversity in 2018:

> "Agricultural biodiversity increases resilience, helps farmers to reduce climatic and economic risks, and can enhance productivity, stability, food security, and nutrition. However, global shifts in food production and dietary patterns are threatening agricultural biodiversity. Today, 30 crops supply 95% of the calories that people obtain from food, and only 4 crops—maize, rice, wheat, and potatoes—supply over 60% of those calories. This growing reliance on an increasingly narrow range of crop varieties undermines the ability of agriculture to adapt to climate change because many local crop varieties and animal breeds are more resilient than the modern ones that are replacing them."

Furthermore, a recent report from the FAO has identified industrial agriculture as a major cause of the water crisis. It showed that chemical-intensive industrial agriculture simultaneously demands more water and destroys the soil's water-holding capacity, depleting and polluting 75% of the planet's water and soil. The nitrates in water from industrial farms also create "dead zones" in the oceans.

The Intergovernmental Panel on Biodiversity and Ecosystem Services has warned that land degradation, desertification, and disappearance of biodiversity are already affecting the livelihood and survival of millions. They've found that fertile soil is being lost at a rate of 24 billion tons a year through intensive farming and non-sustainable agriculture, predicting that there could be 700 million refugees if soil, biodiversity, and ecosystems are not regenerated.

As analyzed in *Soil Not Oil*, industrial agriculture is a major contributor to climate change. Nearly 40% of all greenhouse gas emissions responsible for climate change come from the fossil fuel and chemical-intensive industrial system of agriculture. In contrast, biodiverse organic farming contributes to mitigation, adaptation, and resilience.

Public Health

Not only has industrial agriculture damaged planetary health, but it has also undermined the right to food and negatively impacted the health of people. Around 75% of chronic diseases have their roots in the food we eat and the toxins in the environment.

While the destruction of biodiversity and ecological capital is justified in terms of "feeding people," the problem of hunger has grown. Over 1 billion people are consistently hungry. Another 2 billion suffer from food-related diseases such as obesity. Hunger and malnutrition are designed into a food system driven by profits rather than health and sustainability.

As touched upon earlier, this system has its origins in making explosives and chemicals for war, which later remodeled itself as the agrochemical industry when those wars ended. Explosive factories started to make synthetic fertilizers, and war chemicals started to be used as pesticides and herbicides. In 1984, a gas leak from a pesticide plant led to the Bhopal disaster (also known as the Bhopal gas tragedy), where roughly 3,000–7,000 people died in the immediate aftermath. Since then, the calamity has killed over 15,000 more and left hundreds of thousands of Bhopali residents with long-term health conditions. This is a stark reminder that pesticides kill. The UN report on "Pesticides and Right to Food" states that:

> "Pesticides cause an estimated 200,000 acute poisoning deaths each year, 99% of which occur in developing countries. Hazardous pesticides impose large costs on governments. Harmful insecticides have catastrophic impacts on health and the potential for human rights abuses against farmers and agricultural workers, communities living near agricultural lands, indigenous communities, and pregnant women and children." (UNGA 2017)

The Navdanya book *Poisons in Our Food* highlights a further link between disease epidemics like cancer and the use of pesticides in agriculture, suggesting that a daily "cancer train" leaves Punjab—the land of the Green Revolution in India—with cancer victims. Nearly 33,000 people have died of cancer in Punjab in the last five years. Andre Leu's books *The Myth of Safe Pesticides* and *Poisoning Our Children* also document the harm to human health from pesticides.

Farmers' Livelihoods and Rural Economies

A capital-intensive external input system has also increased the costs of production where farmers spend more than they earn, consequently trapping farmers in debt. Industrial agriculture combined with the globalization of trade in food has created an unprecedented farming crisis, triggering an epidemic of farmer suicides across the world. We have the choice to farm in ways that create abundance, are economically viable, and result in prosperous outcomes for farmers.

A flawed economic paradigm was created to promote industrial agriculture, which presented a negative economy as "productive" and necessary for feeding the world. Industrial farming uses ten times more energy as input than it produces in food. It also uses higher financial inputs than the farmer can recover, given the collapsing prices of globally traded commodities, leading to a debt trap and displacement. The pseudo-productivity hides true costs of damage to be borne by the Earth and society. These hidden externalities are the basis of the ecological, farmer, and health crises that are associated with industrial agriculture. On the basis of the myth of productivity, agriculture became focused on large industrial farms producing chemical monocultures of a handful of commodities. As costs of production increased, farmers were trapped in debt, and small farmers started to disappear. Landholdings were consolidated, not because large farms are more efficient, but because they get most of the subsidies. The total agricultural subsidies are $500 billion, favoring large-scale farms. In terms of resource use and productivity, small farms are more productive. A former Prime Minister of India, Charan Singh, recognized this when he wrote:

> "Agriculture being a life process, in actual practice, under given conditions, yields per acre decline as the size of farm increases (in other words, as the application of human labor and supervision per acre decreases). The above results are well-nigh universal: output per acre of investment is higher on small farms than on large farms. Thus, if a crowded, capital-scarce country like India has a choice between a single 100-acre farm and forty 2.5-acre farms, the capital cost to the national economy will be less if the country chooses the small farms."

The negative impact on small farmers has also had an impact on the health and nutrition of people. The human diet has shifted from 8,500 plant species to about eight globally traded commodities that are nutritionally empty. As suggested earlier, the destruction of biodiverse small farms has had a significant impact on health, leading to the growing epidemic of non-communicable chronic diseases such as cancer, cardiovascular diseases, hypertension, neurological problems, intestinal problems, and infertility.

When the focus is the production of commodities for trade instead of nourishment, disease and malnutrition are the outcomes. Only 10% of corn and soy grown is used as food—the rest goes for animal feed and biofuel. Commodities do not feed people; foods rich in nutrients do. Furthermore, "cheap" commodities have a very high cost financially, ecologically, and socially. These commodities are artificially kept afloat with $500 billion in subsidies (more than $1 billion a day), creating massive debt. Debt and mortgages are the main reason for the disappearance of the family farm. In extreme cases, especially in the cotton belt of India, debt created by the purchase of expensive seeds and chemical inputs has pushed more than 300,000 farmers to suicide in a little over two decades. Getting out of this suicide economy has become urgent for the well-being of farmers, eaters, and all life on earth.

Instead of an ecological approach based on interconnectedness, agriculture has become compartmentalized into fragmented disciplines based on a reductionist, mechanistic paradigm. Instead of focusing on the ecological functions of biodiversity in the soil and among plants, animals, and insects, agriculture has been reduced to external inputs of chemical fertilizers, pesticides, fungicides, and herbicides.

Just as GDP fails to measure the real economy, the health of nature, and society more broadly, the category of "yield" fails to measure the real costs of external inputs and the real output of farming systems. As the UN observed, the so-called High Yielding Varieties (HYVs) of the Green Revolution should, in fact, be called High Response Varieties since they are bred for responding to chemicals and are not high yielding in and of themselves. The narrow measure of "yield" propelled agriculture into deepening monocultures—which displaces diversity and destroys biodiversity's ecological services and functions—thereby eroding natural and social capital. Within the industrial model of agriculture, it is impossible for India and other countries to meet the Sustainable Development Goals they have committed themselves to.

1.3 *A Biodiversity-Based Approach to Farming: Agroecology and Regenerative Organic Agriculture*

We need a new paradigm of working with the laws of nature and ecological sustainability, not against them. Nature's laws are based on biodiversity and agroecology. Industrial agriculture is based on external inputs of chemicals, which destroys biodiversity and its ecological functions. There is a strong consensus that the main agricultural production systems being used to produce the world's food must change because they are clearly unsustainable. Many experts also agree that the current knowledge base used to underpin conventional agriculture is not sufficient to do this:

> "The formal AKST system [agricultural knowledge, science, and technology] is not well equipped to promote the transition toward sustainability. Current ways of organizing technology generation and diffusion will be increasingly inadequate to address emerging environmental challenges, the multi-functionality of agriculture, the loss of biodiversity, and climate change. Focusing AKST systems and actors

on sustainability require a new approach and worldview to guide the development of knowledge, science, and technology, as well as the policies and institutional changes to enable their sustainability. It also requires a new approach in the knowledge base." (IAASTD 2008)

A scientifically and ecologically robust paradigm of agriculture is emerging based on the paradigm of agroecology—the science of ecology applied in agriculture. Instead of chemical inputs which cause harm to the environment and public health, the ecological agriculture paradigm is based on biodiversity. Regenerative organic farming is beneficial to the soil, water, climate systems, public health, and farmers' livelihoods. Agroecology puts biodiversity at the heart of food production. It changes the measure of productivity from yields of monoculture commodities produced with intensive fossil fuels and chemical inputs to the biodiversity-based total output of biodiverse systems, including the internal input ecological functions provided by biodiversity, which are alternatives to chemical inputs.

Navdanya's practice and research of three decades have shown that we can regenerate biodiversity, soil, and water, as well as mitigate climate change, increase nutrition and health, and double food production and farmers' incomes through the utilization of biodiverse, regenerative ecological agriculture. Since the 1980s, Vandana Shiva has been practicing and promoting non-violent biodiverse agriculture. She realized that the term "The Monoculture of the Mind" is a terminology defining the framework that prioritizes "yield" (only a small part of biodiverse ecosystems), which proposes that chemical farming outputs increase overall production and is, therefore, the solution to food insecurity.

Through Navdanya, Shiva started to look at biodiversity-based productivity and found the total output to be much higher than the monoculture yields of chemical farming. Navdanya started to measure health per acre and nutrition per acre rather than yield per acre (Navdanya, *Health Per Acre*, 2011). Based on this research, it was found that biodiverse-intensive organic farming can feed twice the population of India while conserving our natural resource base.

The FAO has also reiterated the link between biodiversity and diets stated in its press release for the International Day for Biological Diversity. It is now recognized that biodiversity in our fields is connected to biodiversity in our diets. As Navdanya's research on biodiverse organic systems has shown, ecological systems produce higher outputs and incomes for rural families. The Navdanya report *Health Per Acre* shows, when measured in terms of nutrition per acre, ecological systems produce more food. Biodiversity-based organic agriculture also reduces farmers' costs by using the multifunctional ecological principles of agroecology. The Navdanya book on true-cost accounting, *Wealth Per Acre*, shows how biodiversity and agroecology are an answer to rural (including farmers') poverty and the agrarian crisis. Ecological systems of agriculture are based on care, compassion, and cooperation to enhance ecological resilience and diversity, sustainable livelihoods, and health.

This new paradigm of agriculture creates living economies and living cultures, which increases the well-being of all people and all beings. At the heart of this system are biodiversity and agroecology, both as a paradigm and as a means of production. As this work with Navdanya and many organizations across the globe shows, we can produce more nutrition and higher incomes for farmers through biodiversity-based organic and regenerative farming, which regenerates the planet's soil, biodiversity, water, climate systems, health, farmers' livelihoods, and food democracy.

Biodiversity

Born in the forests of the Himalayas, Shiva has walked the diversity way since her childhood. In the 1970s, when she became a volunteer in the Chipko movement, the contrast between the two paradigms of forestry—one based on commerce and monocultures, the other based on sustenance and diversity—became stark. From her sisters (none of whom had ever been to university), she learned lessons of biodiversity and interconnectedness; how the forest was connected to streams and rivers on the one hand and to sustainable agriculture on the other. The slogan of the women was that the primary products of the forests were not timber, resin, and revenue but soil, water, and pure air. In 1981, the Chipko movement was successful in getting a logging ban in the high Himalaya, and since then, mountain forests have been recognized for their ecological functions of soil and water conservation, along with the renewal of clean air. Shiva's education in the nutrient and hydrological cycles took place in what she calls the "Chipko University of Biodiversity"—even while she was doing her PhD on "Hidden Variables and Non-locality in Quantum Theory" at the University of Western Ontario in Canada.

In 1982, Shiva was asked by the United Nations University to undertake a five-year study on Conflicts Over Natural Resources. In 1984, as part of the UNU study, her attention was drawn to the tragedies of Punjab and Bhopal, and she completed what was later published as the book, *The Violence of the Green Revolution.*

Her biodiversity journey in agriculture began with trying to understand the violence built into chemical farming. Blindness to biodiversity and its ecological functions is central to introducing chemicals that harm the Earth, biodiversity, and health. Chemical agriculture has made us forget the role that biodiversity plays in sustainable agriculture.

Biodiversity represents the variety of plants, animals, and microorganisms in the world, along with their ecological functions and the relationships between them. The higher the diversity and the more multidimensional its ecological functions, the more stable and sustainable a system is, and the higher the goods and services it can provide.

Biodiversity of seed and plant varieties is necessary for increasing soil health, water conservation, and carbon sequestration. The diversity of ecological functions performed by biodiversity renews soil fertility and contributes to pest and weed management. The alternative to chemical fertilizers, pesticides, and herbicides that are harming the health of the planet and people is a biodiversity of plants, insects, birds, soil organisms, and farm animals.

The level of interactions between various biotic and abiotic components determines the overall behavior of an agroecosystem that translates into agricultural performance. Increasing levels of functional biodiversity within an agroecosystem triggers a phenomena of synergisms that ameliorate soil biology, recycle nutrients, enhance photosynthetic efficiency, and provide other biological functions. In other words, a high level of biodiversity puts the entire agroecosystem in a state of enhanced dynamism. The more dynamic an agroecosystem, the more functional, productive, and sustainable it can become.

Traditional farmers are replete with the wisdom of biodiversity. They have evolved various tactics of enhancing biodiversity at every level in their farming systems. Due to this fact, traditional agriculture has not only survived over millennia, but also performed well and sustainably. Reverberating with lively biodiversity—traditional agriculture is eternal. Ever-enhancing biodiversity is the crux of evolution, and traditional farmers have always understood this reality and articulated it in their farming strategies. Every innovation of farmers revolves around biodiversity. Their discovery of new varieties and characterization of each one is a wonderful way of enriching agriculture with new experiences.

The erosion of agrobiodiversity led to the erosion of traditional agriculture. Attempts were made by the proponents of industrial agriculture to defame traditional agriculture. In the meantime, when most landraces of several food crops vanished—perhaps forever—traditional agriculture lost much of its appeal. It remained confined only to a few isolated and poor areas where genetic erosion could not take place.

However, biodiversity in agricultural practices can be restored in several ways following the principles of agroecology. Many scholars have stressed the key importance of biodiversity and suggested ways to restore biodiversity in agriculture. The following are a few strategies of agroecosystem diversification:

Agroforestry Systems: Trees or other woody perennials, annual crops, and livestock are integrated to enhance complementary relations between components, increasing multiple uses of the agroecosystems.

a) **Polycultures:** Two or more crop species are planted together. For example, planting shallow-rooted millets with deep-rooted pulses so that a higher yield of more than one food (economic product) is taken per unit area.

b) **Cover Crop:** Plant pure or mixed stands of legumes or other annuals under fruit trees for improving soil fertility, creating biological control of pests, and modifying the microclimate.

c) **Crop Rotations:** Temporal diversity in cropping systems provides nutrients and breaks life cycles of several insect pests, diseases, and weeds.

d) **Livestock:** Livestock husbandry in agroecosystems creates extra nutrient and energy pathways to enable the production of a variety of foods and enhance nutrient recycling. All the diversified forms of agroecosystems share the following features (Altieri, 2000):

  • Vegetative cover is maintained as an effective soil and water-conserving measure met through the use of no-till practices, mulch farming, use of cover crops, and other appropriate methods.

  • They are provided a regular supply of organic matter through the addition of organic matter (manure, compost, and promotion of soil biotic activity).

- Nutrient recycling mechanisms are enhanced using livestock systems based on legumes and other cover crops.
- Pest regulation is promoted through enhanced activity of biological control agents achieved by introducing or conserving natural enemies and antagonists.

Biodiversity is at the heart of our approach to designing farming systems based on agroecological principles. In agroecosystems, biodiversity is of critical value for a variety of reasons (Altieri 1994, Gliessman 1998):

- As diversity increases, so do opportunities for co-existence and beneficial interactions between species that can enhance agroecosystem sustainability.
- Greater diversity allows for more efficient resource use in agroecosystems through better system-level adaptation to habitat heterogeneity. This leads to complementarity in crop species' needs, diversification of niches, overlap of species niches, and partitioning of resources.
- In diverse ecosystems where plant species are intermingled, there is a greater abundance and diversity of pests' natural enemies, which keeps populations of herbivore species in check.
- A diverse crop assemblage can create a diversity of microclimates within the cropping system that can be occupied by a range of non-crop organisms—including beneficial predators, parasites, pollinators, soil fauna, and antagonists—that are of importance for the entire system.
- Diversity in the soil performs a variety of ecological services such as nutrient cycling, detoxification of noxious chemicals, and regulation of plant growth; contributes to the conservation of biodiversity in surrounding natural ecosystems; and reduces risk for farmers in marginal areas with unpredictable environmental conditions.

The United Nations Convention on Biological Diversity was signed by 150 government leaders at the 1992 Rio Earth Summit to promote the conservation of biodiversity, its sustainable use, and equitable sharing. Since the ecological functions of biodiversity include provisioning of fresh air, food, water, medicine, and shelter, its conservation is linked to the basic needs of people. Biodiversity-based agroecology is vital to food and nutritional security.

The Global Erosion of Biodiversity

Diversity is a core characteristic of nature and the basis of ecological stability. Diverse ecosystems give rise to diverse life forms and to diverse cultures, and this provides the basis of sustainability. The co-evolution of cultures, life forms, and habitats has conserved the biological diversity on this planet; cultural diversity and biological diversity go hand in hand.

Communities everywhere in the world have developed knowledge and found ways to create livelihoods from the bounties of nature's diversity—in wild and domesticated forms. Hunting and gathering communities use thousands of plants and animals for food, medicine, and shelter. Pastoral, peasant, and fishing communities have also evolved knowledge and skills to derive sustainable livelihoods from the living diversity of the land and water. The deep and sophisticated ecological knowledge of biodiversity has given rise to cultural rules for conservation reflected in notions of sacredness and taboos.

Today, however, the diversity of ecosystems, life forms, and ways of life of many communities are under threat of extinction. Habitats have been enclosed or destroyed, diversity has declined, and livelihoods derived from biodiversity are threatened.

Tropical, moist forests cover only 7% of the Earth's land surface but contain at least half of the Earth's species. Deforestation in these regions is continuing at a rapid pace, with conservative estimates suggesting net rates are as high as 6.5% in Cote d'Ivoire and average about 6% per year (about 7.3 million ha) for all tropical countries. At this rate, incorporating reforestation and natural growth, all closed tropical forests would be cleared within 177 years (FAO 1981). About 48% of the world's plant species occur in or around forest areas where more than 90% of their area will be destroyed during the next 20 years, leading to roughly a quarter of those species being lost (Raven 1988). The current extinction rate is estimated to be about a thousand species a year (Wilson 1988). By the 1990s, this figure was expected to rise to ten thousand species a year, which is one species every hour. During the next 30 years, one million species could be erased.

Biological diversity in marine ecosystems is also remarkable, and coral reefs are sometimes compared with tropical forests in terms of diversity (Connell 1978). Sadly, marine habitats and marine life are under severe threats as well. With the destruction of diversity, the base of fisheries in most coastal regions of the world are on the verge of collapse.

The Growth of Ecological Vulnerability

The erosion of diversity is also severe in agricultural ecosystems. These ecosystems, particularly in the tropical belt, have from time immemorial been the fountainhead of the world's food. Wheat, rice, potatoes, vegetables, and fruits have been dispersed all over the world from these origins of diversity. The landraces of this belt include varieties resistant to drought and pests, medicinal plants, and sources for humankind's shelter and clothing. Furthermore, the varieties most resistant to locally-occurring pests and diseases are Indigenous ones.

Even if certain diseases occur, some of the strains may be susceptible while others have the resistance to survive. Traditionally, cropping patterns like crop rotation have also helped pest control. Since many pests are specific to particular plants, planting different crops in different seasons during different years causes large reductions in pest populations, reducing the need for large chemical applications. They also require less intensive irrigation—a detrimental practice which increases the spread of pests and causes damage to many plant varieties. Such cropping systems thus have in-built protection.

Traditional agriculture does not recognize the concept of "weeds." All plants have their uses; some plants have more than one use. In some cases, the various parts of a single plant each have a separate use. Denial Querol writes about a farm in Mexico which had over two hundred "weeds." The farmer, however, had a specific use for all these plants. To him, not one was a weed fit for destruction.

The deployment of "miracle seeds" of the Green Revolution introduced the concept that only one product of the plant is useful: the marketable one. The High Yield Varieties (HYVs)—or the high response varieties—considered only the grain as the useful product. Plants which did not produce enough grain were regarded as weeds and destroyed even if they were meeting the farmer's needs of fodder, roofing material (as in the case of rice varieties), and other resources. Roughly 30,000 Indigenous varieties of rice grew in India prior to the Green Revolution, while today, there are no more than 50 varieties.

Uniformity meant that fields could not be used to grow more than one crop at a time. Where the farmers either grew chickpeas along with native wheat varieties, or mustard with ragi, uniformity demanded that such practices be eliminated. Thus, sufficiency in food (besides grain) disappeared along with varieties like bathua, amaranth, and other traditionally significant plants. Since only grain was important, it became imperative to not let the fields lie fallow, but to plant the grain repeatedly, leading to buildups of pests in the soil:

> "The introduction of high yielding varieties has brought about a marked change in the status of insect pests... Most of the HYVs released so far are susceptible to major pests with a crop loss of 30 to 100 percent." (Shiva 2016)

The HYVs yielded large amounts of grain only if they were given inordinate amounts of chemicals in the form of fertilizers and pesticides. The need for irrigation also increased, which has led to extreme toxicity and water logging of the soil.

Such uniformity gives rise to monocultures, which are not ecologically sustainable. By definition, monoculture produces identical plants. Thus, if one plant is susceptible, then all are. In 1970–71, America's vast corn belt was devastated by the Southern Corn Blight. Asia's rice was ravaged by bacterial blight in 1968–69, and by tungro disease in 1970–71. Over 2 million acres of Indonesia's rice-producing areas were attacked by pests in 1975. As late as 1992–93, northern India's potato crop suffered severe losses, particularly in Uttar Pradesh.

The increasing vulnerability of agriculture is reflected in livestock farming. Traditional varieties of cows are giving way to crossbreeds of Jersey and Holstein, which are far more vulnerable to diseases and require more planned feeding strategies than Indigenous breeds (which could forage for themselves). These crossbreeds do not produce enough manure—the small farmer's main organic fertilizer—but are merely milk producers.

The erosion of biodiversity starts a chain reaction. The disappearance of a species is related to the extinction of innumerable other species interrelated through food webs and food chains (about which humans do not fully understand). The displacement of Indigenous varieties, whether in crops or in livestock, has serious ecological consequences which undermine productivity. The crisis of biodiversity is not simply just the disappearance of species which generate money for corporate enterprises as industrial raw material. It is, more fundamentally, a crisis that threatens the life-support systems and livelihoods of millions of people in developing nations.

Biodiversity is the primary means of production of sustainable, small-scale agriculture. The ecological functions of biodiversity provide the internal inputs, which allow the farmer to become free of dependence on external chemical inputs such as fertilizers, pesticides, and herbicides. The Navdanya approach to conservation of biodiversity in agriculture is based on conservation of diversity at five levels:

1. Ecosystem diversity
2. Farming system diversity
3. Species diversity
4. Varietal or genetic diversity
5. Output diversity

India is a large country with high diversity in ecosystems. Ecosystem diversity leads to diverse farming systems in which land, water, and biodiversity are managed in different ways with various linkages between livestock, trees, and crops. These systems are built on the ecosystem and biodiversity knowledge that farming communities have developed and refined over generations. Communities have also evolved conservation strategies to ensure the sustainable utilization of natural resources.

Despite biodiversity erosion, the continued existence of species and variety diversity that can be found results from the millions of unknown and invisible farmers' cultural practices and knowledge systems. Most species in Indian farming systems satisfy more than one need or function. Conservation of the farm's diversity is carried out with the understanding that single output measurements distort the full potential of farming systems that utilize different species with multiple yields.

The conservation of diversity on multiple levels is an efficiency and productivity imperative because it allows ecological intensification of agricultural production. Through diversity, resource prudence is created. Such conservation creates jobs because it protects farmers and increases their outputs.

Biodiversity is the Foundation of Agroecology

Agroecology is the scientific paradigm for sustainable agriculture. Agriculture is and should indispensably be a life-enhancing phenomenon. Production of a variety of healthy and nutritious foods requires a productive and healthy agroecosystem reverberating with biodiversity in its forest, cropland,

and livestock. Agriculture based on healthy, biodiversity-laden, and vibrant agroecosystem is naturally the agriculture rooted into its inexhaustible source of nature: the solar-powered agroecosystem.

Agroecology is the holistic study of agroecosystems, including all environmental and human elements. It focuses on the form, dynamics, and functions of their interrelationships and the processes in which they are involved (Altieri 1987; Reijntjes et al. 1992). Intercropping, agroforestry, and other traditional methods mimic natural ecological processes. The sustainability of many local practices lies in the ecological models that agroecologists follow. By designing farming systems that mimic nature, farmers can get the optimal use out of sunlight, soil nutrients, and rainfall (Reijntjes 1992).

Agroecology gives deeper meaning to agriculture. It integrates agriculture with ecology. It helps us understand the direct relationship between agriculture and ecology. It teaches us to be in tune with nature while producing a diversity of healthy, nutritious, and delicious foods using sources of nature. In essence, agroecology is the philosophy of relishing all edibles that nature produces and, at the same time, nurturing nature so that it can blossom with biodiversity.

Agroecology is now a separate discipline of agriculture and ecology. It is the central concept of many valuable ideas, philosophies, approaches, strategies, and tactics of life which include natural farming, traditional agriculture, permaculture, biodynamic farming, integrated pest management, organic agriculture, and sustainable agriculture. Agroecology uses ecological theory to study, design, manage, and evaluate food production systems. It is the concept on which sustainable agriculture—which ensures the future of agriculture— has been built. It is through applying the principles of agroecology that we protect, conserve, and augment natural resources such as forests, grasslands, livestock, soil, water resources, and farming. Agroecology appreciates and strengthens interactions among all crucial biophysical, socioeconomic and technical components of the agroecosystems. All components are regarded to be fundamental units of an integrated system.

Agroecology helps us understand and maintain vital mineral cycles, biological processes, energy transformations, and socioeconomic relationships in an integrated manner. Agricultural strategies woven around the principles of agroecology look into local geographical, socioeconomic, environmental, and cultural specificities and obey traditions, such as food habits, festivities, and ethical or aesthetic values.

A one-dimensional monoculture view of conventional agriculture has no place in agroecology. An understanding of ecological and social levels of co-evolution, structure, and function is instead necessary (Altieri 2000). Rather than focusing on one particular component of the agroecosystem, agroecology emphasizes the interrelatedness of all components and the complex dynamics of ecological processes (Vandermeer 1995). Agroecology is a holistic response to agribusiness-based exploitative technologies and trade for profits, with have no room for other values of life and are not conscious of the future of the planet. Agroecology, on the other hand, does not overlook technical and economic aspects but is very much alive to social, cultural, and environmental issues, firmly standing for the present and future well-being of society.

Food production needs are central to the concept of agroecology. The performance criteria in agroecology takes into consideration vital contemporary issues, namely, ecological sustainability, food security, and climate change mitigation and adaptation. Traditional concepts of organic farming, natural farming, and ecological farming offer to resolve numerous issues from the individual family to the global level, from seed to *swaraj* (self-rule), from agribusiness empire to genuine socialism, from food security to food sovereignty, from ecological disaster to ecological affluence, and from climate chaos to climate order.

1.4 *The Principles of Agroecology*

Principles of agroecology revolve around the three functional biotic components in an agroecosystem: a community of plants, animals, and microorganisms interacting among themselves and with the physical-chemical environment modified by farmers to produce foods, fodder, fiber, fuel, and other useful products. Agroecology provides us an opportunity to have a holistic understanding of the agroecosystems that we design and manage for food production. The design of agroecosystems is based on the following agroecological principles (Reijntjes et al. 1992; Altieri 1987, 2000; Singh 2005):

- Enhancing the recycling of biomass, optimizing nutrient availability, and balancing nutrient flow
- Securing favorable soil conditions for plant growth by managing organic matter and enhancing soil biotic activity

+ Minimizing losses due to flows of solar radiation, air, and water through microclimate management, water harvesting, and soil management through increased soil cover
+ Species and genetic diversification of the agroecosystem
+ Increasing the beneficial biological interactions and synergism among agrobiodiversity components that result in the promotion of key ecological processes and services

Depending on local opportunities, resources constraints, and market needs, various techniques and strategies to influence the productivity, stability, and resilience within an agroecosystem must be applied. The ultimate goal of agroecological design is to:

+ Integrate components so that overall biological efficiency is improved.
+ Preserve biodiversity
+ Maintain agroecosystem productivity and self-sustaining capacity
+ Design a range of agroecosystems within a landscape unit, each mimicking the structure and function of natural ecosystems

Since Altieri and Reijntej elaborated the agroecological principles, much water has flown into Ganga. Now new goals need to be set and added to their vision. The ultimate goal of designing and managing agroecosystems, in view of the changing circumstances, would be to:

+ Enhance ecological integrity for food production sustainability
+ Create socioeconomic and cultural environment for food sovereignty to prevail
+ Build up the microclimate
+ Increase carbon sequestration to effectively deal with climate change

Enhancing biodiversity in agroecosystems biodiversity is the basis of ecological processes. It is biodiversity by which an agroecosystem functions. It is the active principle of an ecosystem's performance, and the root cause of functioning of the biosphere. Less diverse ecosystems, or monoculture-based agroecosystems, are extremely vulnerable and unsustainable. As the biodiversity within an ecosystem enhances, the level of its resilience and sustainability increases. Extreme diversity leads to extreme resilience and the highest level of sustainability. From a management perspective, the agroecological objective is to provide balanced environments, sustain yields, biologically mediate soil fertility, and create natural pest regulation through the designs of diversified agroecosystems and the use of low-input technologies (Gliessman 1998; Altieri 2000). By designing farming systems that mimic nature, optimal use can be made of sunlight, soil nutrients, and rainfall (Pretty 1994; Altieri 2000).

Agroecology and Sustainable Agriculture

No system can be sustainable without ecological integrity. The Green Revolution and biotechnology-based LPG agriculture are not sustainable as they ignore ecological sustainability. Agroecology, on the contrary, relies on ecological sustainability. Ecology, in fact, is the very essence of sustainable agriculture—the type of agriculture, which fulfills the need of the present without compromising with its ability to fulfill the needs of the future. A sustainable agriculture designed in accordance with the agroecology principles would be the one having the following basic characteristics:

1. **Ecologically vibrant:** Embracing the highest level of biodiversity in all its components, it has a larger forest-cultivated land ratio and manages cyclic nutrient flows. It is energy-efficient, regenerative, resource-conservation oriented, enhances carbon sequestration, aids in climate regulation, and has high resiliency.

2. **Economically viable:** It is productive to its maximum potential on sustained basis, meets all necessary requirements of life apart from food and nutrition security, such as education of children of a family, and other day-to-day domestic needs. Most, if not all, of the inputs are produced within the system. Ratio of output-input values must be large enough, i.e., prices of the outputs be higher than those of inputs. It also means it is exporting, rather than importing, in nature.

3. **Socio-culturally:** It takes care of farming communities and of all people, rather than private interests. Food resources and nutrition security are both accessible and ensured for all. It respects local food habits and cultural diversity, recognizes and promotes traditional ecological knowledge, and is governed by farming communities, serving to integrate societies by creating cohesion among people.

Ecological vibrancy, economic viability, and socio-cultural justice are the key traits of sustainable agriculture. Keeping these traits in mind, and applying the principles of agroecology, sustainability can be operationalized through the following methods:

1. **Maintaining an ecological balance by:**
   - Having a large forest to cultivated land ratio
   - Giving full protection to forest area

2. **Enhancing ecological processes by:**
   - Strengthening the immune system (proper functioning of natural pest control
   - Decreasing toxicity through elimination of agrochemicals
   - Optimizing metabolic function (organic matter decomposition and nutrient cycling)
   - Balancing regulatory systems (nutrient cycles, water balance, energy flow, population regulation, etc.)
   - Enhancing conservation and regenerating soil-water resources and biodiversity
   - Increasing and sustaining long-term productivity

3. **Implementing mechanisms to improve agroecosystem resilience by:**
   - Increasing plant species and genetic diversity
   - Enhancing functional biodiversity (natural enemies, antagonists, etc.)
   - Enhancement of soil organic matter and biological activity
   - Increasing soil cover and crop competitive ability
   - Eliminating toxic inputs and residues

4. **Optimizing the use of locally available resources:**
   - Combining the different components of the farm system, i.e., plants, animals, soil, water, climate, and people
   - Supporting the greatest synergetic effects possible

5. **Reduce the use of off-farm, external and non-renewable inputs with the greatest potential to damage the environment or harm farmers and consumers by:**
   - Relying mainly on resources within the agroecosystem by replacing external inputs with nutrient cycling, better conservation, and an expanded use of local resources
   - Improving the match between cropping patterns and the productive potential and environmental constraints of climate and landscape to

ensure long-term sustainability of current production levels
- Working to value and conserve biological diversity, both in the wild and in domesticated landscapes, and making optimal use of biological and genetic potential of plant and animal species; taking full advantage of local knowledge and practices, including innovative approaches not yet fully understood by scientists although widely adopted by farmers

The Cycle of Sustainability

Sustainability is not a static phenomenon, but rather a dynamic phenomenon. Sustainability is also not a final outcome; it is a cyclical process. Traditional farmers manage the soil in such a way that it should continue to be replenished by nutrients through manure, recycling, *in situ* fertilization, mixed cropping, mulching, and other management practices. They still adhere to an old adage: don't feed the plant, feed the soil which feeds the plant.

Farmers cultivate as much agrobiodiversity as could be possible in a particular area. They also manage the natural biodiversity in uncultivated areas (forests, grasslands, rangelands, etc.). This biodiversity is a key to sustainability—the higher the degree of biodiversity, higher the level of sustainability. Farmers also manage cyclic flows of nutrients. Whatever nutrients are extracted from croplands are recycled into the same soil through manure. The soil fertility is further enhanced by supplementing the nutrients from forest soil.

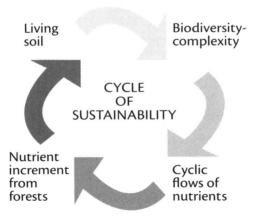

Living soil — Biodiversity-complexity — Cyclic flows of nutrients — Nutrient increment from forests

CYCLE OF SUSTAINABILITY

This wonderful practice of traditional farming is an example of farmers' management of sustainability in traditional agriculture (Fig. 1). Thus, sustainability of agriculture is not a certain static state or the level of production; it is a paradigm and a process, a dynamic phenomenon which completes its cycle fed at each of the four steps: biodiversity-complexity, flows of nutrients, nutrient increment from forests, and the living soil.

Diverse systems of sustainable agriculture practice agroecology under different names, and with slightly different focuses. Permaculture, biodynamic agriculture, natural farming, and organic farming are different schools of practice. The principles of all non-industrial, non-chemical agriculture systems are the principles of agroecology.

The FAO has identified the following "10 Elements of Agroecology" for planning, managing, and evaluating agroecological transitions:

1. Diversity
2. Co-creation of knowledge and transdisciplinary approaches
   for innovation
3. Synergies
4. Efficiency
5. Recycling
6. Resilience
7. Human and social value
8. Culture and food traditions
9. Responsible governance
10. Circular and solidarity economy

We focus on regenerative organic farming because of the three-fold crisis of ecological, societal, and individual health. It focuses on seeds, soil, water, and climate systems to benefit the lives of farmers.

The Four Principles of Organic Agriculture

The four principles of organic agriculture were developed from current organic practices through extensive worldwide consultation by the International Federation of Organic Agricultural Movements (IFOAM). They are the agreed international consensus on the fundamental basis of organic production. These principles are used to inform the development of practices, programs, and standards:

1. Health
2. Ecology
3. Fairness
4. Care

IFOAM is the international umbrella movement that has the role to both lead and unite the organic sectors around the world. It is the organization that sets the international standards, policies, definitions, and positions around the multifunctionality of organic agriculture through consulting with its members that cover the whole spectrum of the sector in the majority of countries around the world. Consequently, IFOAM documents are seen as highly credible source texts for reference material.

The mutually agreed upon definition of organic agriculture that IFOAM has developed clearly shows that organic systems should be based on environmental and social sustainability by working with the ecological sciences, natural cycles, and farmers:

> "Organic agriculture is a production system that sustains the health of soils, ecosystems and people. It relies on ecological processes, biodiversity and cycles adapted to local conditions, rather than the use of inputs with adverse effects. Organic agriculture combines tradition, innovation and science to benefit the shared environment and promote fair relationships and a good quality of life for all involved."

Two Significant Differences Between Conventional and Organic Systems

Most organic production standards clearly state that organic production avoids the use of synthetic fertilizers and pesticides. The United Nations Food and Agriculture Organization (FAO) created the international standard for the trade in food products, *The Codex Alimentarius*, which covers organic production. The *Guidelines for the Production, Processing, Labelling and Marketing of Organically Produced Foods* (acquired by the Codex Alimentarius Commission) states that "Organic agriculture is based on minimizing the use of external inputs, avoiding the use of synthetic fertilizers and pesticides."

This is an important distinction between conventional and organic products. Unfortunately, it has also led to the wrong assumption that because organic systems do not use two key conventional farming inputs: fertilizers to correct nutritional deficiencies and methods to stop pests and diseases. Organic is seen by some authorities as farming by neglect that produces low-yields of inferior produce (Avery 2000; Trewavas 2001). This is not the case, and this book presents key environmental benefits based on credible published literature. Several peer-reviewed comparison studies have found that organic agricultural systems are the most environmentally sustainable and have the least off-farm impact compared to our current agricultural systems (Drinkwater 1998; Welsh 1999; Reganold et al. 2001; Mader et al. 2002; Hole 2004; Pimentel 2005). Environmental sustainability, especially in terms of working with ecological systems, and ensuring healthy ecosystems by actively caring about the production methods, is fundamental to organic agriculture. This is an intentional outcome based on the core principles that underpin organic production systems.

Organic farming is not an external input substitution system. It is not a system of neglect. It negates the need for synthetic pesticides and fertilizers by intensifying biodiversity and its ecological functions for improving soil fertility by using composts, natural minerals, cover crops, and recycling organic materials. Biodiversity and cultural and ecological management systems are used as the primary control of pests, weeds, and disease, with a limited use of natural biocides of mineral, plant, and biological origin as the tools of last resort.

Recycling Organic Matter is the Fundamental Basis of Organic Farming

The key to the health of land-based ecosystems is to ensure that the soil—which can support and produce complex food webs—is continually fed with organic matter. This is the reason J. I. Rodale popularized the term "organic" in the 1940s, and it is the fundamental basis of organic agriculture.

Rodale was inspired by Albert Howard who was sent to India as the "imperial economic botanist" by the British Empire in 1905. When he arrived in India at Pusa in Bihar, he found the soils were fertile and there were no pests in the field. He decided to make the pests and peasants his professors to learn how to farm well. From this experience, he wrote the organic agriculture classic *The Agricultural Testament*, which inspired Rodale. Rodale was the first major international author and publisher of books and magazines on organic farming. His major magazine, *Organic Farming and Gardening*, was widely read by many thousands of people around the world. He actively promoted the phrase "organic farming" in this and other publications. Rodale repeatedly stated that the fundamental basis of organic farming was to improve soil health and build up humus through a variety of practices that recycled organic matter (Rodale 2011).

Ecological regenerative agriculture is based on recycling organic matter, hence recycling nutrients. It is based on the Law of Return and on giving nutrients back to the soil—not simply taking nutrition out of it. According to Albert Howard, taking without giving back to the foundation that sustains our global community's livelihood is "a particularly mean form of banditry, because it involves the robbing of future generations which are not here to defend themselves."

As the ancient Vedas recognized 4,000 years ago, "Upon this handful of soil our survival depends. Care for it, and it will grow our food, our fuel, our shelter and surround us with beauty. Abuse it, and the soil will collapse and die, taking humanity with it."

In living soil lies the prosperity and security of civilization. In the death of soil is the death of civilization. Our future is inseparable from future of the Earth. It is no accident that the word "human" has its roots in *humous* (soil in Latin), and Adam, the first human in Abrahamanic traditions is derived from *Adamus* (soil in Hebrew). We forget that we are soil. In taking care of the soil, we reclaim our humanity and produce more food on less land. Fertile soils are the sustainable answer to food and nutrition security.

Soils rich in living organic carbon and humus are also the most effective solution for mitigation and adaptation to climate change. A fossil fuel driven economy, including industrial agriculture, has increased the concentration of carbon dioxide in the atmosphere to levels which are triggering climate instability. We need to reduce emissions and reduce carbon concentrations in the atmosphere. Organic farming offers an ecological process to take excess carbon from the air, where it does not belong, and puts it in the soil, where it does belong. Soil rich in organic matter also holds more water, reducing the demands for irrigation, increasing drought resistance, and creating climate resilience.

Organic Farming Rejuvenates Biodiversity

Research conducted in Europe and the United States shows that organic systems have the highest biodiversity in the fields of the farm compared to other farming systems (Reganold et al., 2001; Mader et al., 2002; Pimentel 2005).

The largest review of 76 studies from around the world comparing organic to conventional agriculture published in the journal *Biological Conservation* found that organic farming increases biodiversity at every level of the food chain from soil biota (such as bacteria) to higher animals (such as mammals):

> "It identifies a wide range of taxa, including birds and mammals, invertebrates and arable flora that benefit from organic management through increases in abundance and/or species richness. It also highlights three broad management practices (prohibition/reduced use of chemical pesticides and inorganic fertilizers; sympathetic management of non-cropped habitats; and preservation of mixed farming) that are largely intrinsic (but not exclusive) to organic farming, and that are particularly beneficial for farmland wildlife." (Hole et al. 2004)

An earlier report by the Food and Agriculture Organization of the United Nations (FAO) stated: "Organic agriculture has demonstrated its ability to not only produce commodities but also to 'produce' biodiversity at all levels" (FAO 2003).

Endemic Biodiversity

The world is going through the greatest extinction event since the end of the Cretaceous period, which resulted in the loss of dinosaurs. Our current loss of biodiversity is called the Anthropocene Extinction and is primarily due to land clearing for agriculture. (Tilman et al., 2001, MA Report 2005).

This extensive loss of habitat and species is also causing major health and social problems around the planet (Sala 2009). Most of this loss is due to land clearing and habitat destruction for the production of agricultural commodities, primarily oil palm, rubber, soybean, sugar, beef, and timber (ETC Group 2009).

One of the criticisms of organic agriculture is that lower yields mean that more land needs to be cleared to fulfill the projected need for more food globally. Another criticism is that organic certification systems do not prohibit the clearing of old growth and valuable habitats and are, therefore, no better at conserving these ecosystems than conventional farming. The conclusion is that since less land is needed to be cleared for conventional systems, they are better for the environment than organic systems (Avery 2000; Trewavas 2001).

The criticism of the environmental integrity of organic systems is very misleading. Organic systems do not have to be lower yielding. The work of the United Nations and Navdanya over the past three decades shows that biodiversity-intensive ecological agriculture systems produce more food and nutrition per acre, and they are land conserving as opposed to chemical-intensive monoculture systems. Section 6.4 "Towards a Biodiversity-Based Framework" shows that best practice organic systems are high yielding. This is particularly relevant in the tropical regions of the world where most of the global land clearing and habitat loss is occurring. All farmers can be easily taught how to adopt high-yielding organic systems. Where this has been done in Africa, it has led to more than a 100% increase in yields. Training farmers in how to adopt best practice organic systems means that more land can be conserved for biodiversity protection (UNEP-UNCTAD 2008).

The reason why most organic standards do not have specific clauses prohibiting the clearing of old growth and ecologically valuable habitat is because it has never been a problem and therefore, they are silent on the issue. There is no evidence of the widespread clearing of these valuable ecosystems for commercial organic production, so this is clearly a problem that does not exist. The overwhelming majority of habitat destruction is for large-scale commodity production, particularly for genetically-modified soy and palm oil. These productions systems are clearly for conventional agribusiness to produce the raw materials that are used to supply industrial processes—the antithesis of most organic systems—which has nothing to do with producing food to feed the hungry (ETC Group 2009).

Conserving and Valuing Habitats

In a strong contrast to millions of hectares of habitat that are being destroyed by conventional agribusiness, organic certifications systems are conserving millions of hectares of habitat by allowing managers to receive a premium for sustainably harvested wild products. Certified organic wild collection is a new and rapidly growing area that conserves over 41 million hectares around the world (Willer 2011).

This is a very important activity as it allows the sustainable harvest of wild resources, ensuring the conservation of high biodiversity ecosystems, and providing an income to people who manage these ecosystems. It is one of the most successful examples of a market-based system that rewards landholders for ecosystem services and provides an economic incentive to conserve ecosystems so that they remain sustainable sources of income.

The certification system defines the criteria and management requirements to ensure that the products are harvested sustainably and that the habitat is not degraded and is sustainably managed. The majority of the certified organic wild harvest areas are in developing countries and the premium prices that are received for the products brings enormous benefits to the communities that manage these ecosystem.

Farm-Based Agricultural Biodiversity

Organic farmers have always been actively involved in conserving traditional varieties of agricultural plants and animals. Many more varieties tend to be cultivated in organic systems than conventional systems due to its roots being based in traditional agriculture and the need for diverse rotation systems.

Navdanya—which means nine seeds—has evolved as a biodiversity-based model of agriculture which enhances diversity of seeds, plants, insects, pollinators, and soil organisms. Intensifying biodiversity also increases food and nutritional security as our work on health per acre and nutrition per acre has shown in our research throughout this book. The biodiversity-centered paradigm of agroecology is being adopted by the Governments of Northeast India and Madhya Pradesh in the form of nutrition-sensitive agriculture. The advent of industrial agriculture has seen a massive decline in on-farm biodiversity as these commercial systems focus of fewer varieties to concentrate on uniformity in production to supply supermarket chains and brand lines. Research by Pat Mooney and colleagues at the ETC group have identified this continuous decline in the biodiversity used in our industrial farming systems (ETC Group 2009):

> "The industrial food chain focuses on far fewer than 100 breeds of five livestock species. Corporate plant breeders work with 150 crops but focus on barely a dozen. Of the 80,000 commercial plant varieties in the market today, well over half are ornamentals. What remains of our declining fish stocks comes from 336 species accounting for almost two-thirds of the aquatic species we consume."

The majority of the world's farmers are involved in traditional farming systems that clearly fit within the organic umbrella. These farming communities are responsible for conserving an enormous amount of unique farm-based biodiversity. When the ETC group researched the traditional systems that fit within the organic paradigm, they found that:

> "Peasants breed and nurture 40 livestock species and almost 8,000 breeds. Peasants also breed 5,000 domesticated crops and have donated more than 1.9 million plant varieties to the world's gene banks. Peasant fishers harvest and protect more than 15,000 freshwater species."

The loss of these organic farming systems brings with it the extinction of irreplaceable biodiversity that is uniquely adapted to these regions. It is important that these farming systems are promoted in order to preserve this immense and valuable biodiversity.

Soil Biodiversity

Nutritive soil contains the highest levels of biodiversity in the planet, forming the soil food web, which is the fundamental basis of most terrestrial ecosystems. Research done by microbiologists for more than 100 years is starting to develop an integrated understanding of the principle roles of key species and how they are important to the nutrient and health cycles of higher organisms. The biodiversity of the soil is so complex and diverse that most of the species are not described by science, and their specific environmental roles are not fully understood (Bardgettr 2005). The vast majority of soil biodiversity is found in the soil organic matter (SOM), especially around the organic matter exudates that are formed by the roots of living plants (Stevenson 1998). Soil biodiversity is primarily fed by the products of photosynthesis from plants and photosynthetic microorganisms such as cyanobacteria. Photosynthesis uses solar power to combine carbon dioxide and water to produce oxygen and two simple sugars: glucose and fructose. Glucose is the basic molecule of life; it is the energy source of the cells of plants and animals. Glucose molecules can be combined and slightly modified to build numerous other types of sugars that life uses, such as sucrose (cane sugar), dextrose (fruit sugar), and lactose (milk sugar).

Glucose molecules can be combined together in long chains to form cellulose. Cellulose is the primary polymer that forms the trunks, branches, and leaves of plants. It is used as paper and timber when processed. Glucose molecules can be also combined together into different types of long chains to form carbohydrates. These are the starches that we eat in our grains, flours, potatoes, and other staple food products. Carbohydrates are modified by plants and animals to form hydrocarbons. These are the oils and fats in our diets.

The energy in gas, coal, and oil that we use to power our modern industries come from fossil fuels produced during photosynthesis millions of years ago. Most living organisms modify carbohydrates with the addition of nitrogen and sometimes sulfur to form amino acids. These are the basis of proteins, DNA, hormones, and other essential molecules of life. Nearly all life on earth is dependent on the products of photosynthesis, either directly or indirectly, as in the case of microorganisms and animals. The complex soil biodiversity recycles all of these organic molecules created initially through photosynthesis by degrading organic matter and through the synthesis of new compounds to build new organic matter. These complex cycles of building, decay, and rebuilding of organic molecules are the fundamental basis of all soil systems and the higher forms of life, including humans who ultimately feed on its products.

Higher Levels of Soil Organic Matter

The main reasons why organic farms have higher levels of soil biodiversity are because of the higher levels of organic matter and the avoidance of synthetic chemicals. Generally, the higher the level of soil organic matter (SOM), the higher the level of soil biodiversity (Stevenson 1998, Zimmer 2000).

Two independent global meta-analyses have reviewed comparison studies between organic and conventional farming systems to determine if there are differences in the rate increase in soil carbon. The study by FiBL and the UK Soil Association has found that organic farming practices sequester around 4,409 lbs of carbon dioxide from the air each year in a hectare of farmland. Both studies included data from Australian studies (Azeez 2009, Gattinger 2011).

On the other hand, most studies of conventional systems find either declines or at best very small increases. Extensive work by scientists at CSIRO has found that traditional Australian faming systems mostly lost soil carbon:

"A major conclusion that can be drawn from this compilation of Australian field trial data (Table 3) is that when SOC [soil organic carbon] stocks were followed through time, even the improved management often showed significant declines, which, in many cases, was likely a direct result of these soils still responding to the initial cultivation of the native soil (e.g. Fig. 7).

However, since the traditional management practice often lost SOC at a greater rate, when only comparing the two treatments at the end of the trial there was a relative SOC gain in the improved management treatment. This means that, at least for the more traditional agronomic systems tested in these trials, Australian soils will generally only be mitigating losses and not actually sequestering additional atmospheric CO_2." (Sanderman, 2010)

Synthetic Nitrogen Fertilizers Degrade Soil Carbon

One of the main reasons for the differences in soil carbon between organic and conventional systems is that synthetic nitrogen fertilizers degrade soil carbon. Research shows a direct link between the application of synthetic nitrogenous fertilizers and the decline in soil carbon. According to La Salle and Hepperly:

"The application of soluble nitrogen fertilizers...stimulates more rapid and complete decay of organic matter, sending carbon into the atmosphere instead of retaining it in the soil as the organic systems do." (La Salle and Hepperly 2008)

Scientists from the University of Illinois analyzed the results of a 50-year agricultural trial and found that synthetic nitrogen fertilizer resulted in all carbon residues from the crop disappearing as well as an average loss of around 22,046 lbs of soil carbon per hectare. This is around 80,909 lbs of carbon dioxide per hectare on top of the many thousands of pounds of crop residue that is converted into CO_2 every year (Khan et al. 2007; Mulvaney et al. 2009).

The researchers found that the higher the application of synthetic nitrogen fertilizer, the greater the amount of soil carbon lost as CO_2. This is one of the major reasons why conventional agricultural systems have a decline in soil carbon while organic systems increase soil carbon. There is a significant body of peer-reviewed evidence showing how the pesticides, herbicides, and fungicides used to kill pests, diseases, and weeds are also toxic to the soil biology. Rachel Carson wrote about this in 1962 and this evidence continues to increase (Carson 1962; Colborn 1996; Cadbury 1997).

These changes to the soil biota create reductions in the production of soil nitrogen, phosphorous, and other plant available nutrients. They also see crop losses due to plant pathogens such as fungal diseases, bacterial wilts, and other diseases. (Cox 2001; Cox 2004; Huber 2010).

Erosion and Soil Loss

Soil loss and erosion in farming systems are major concerns around the world and are major reasons for losses in productivity. Soil loss is also one of the major contributors to eutrophication of aquatic systems, which causes toxic algal blooms, fish kills and the deaths of corals, sea grasses, planktons, daphnias, and numerous marine and freshwater aquatic species due to the loss of oxygen and sunlight due to increases in turbidity (MA Report 2005; IAASTD 2008). Comparison studies have shown that organic systems have less soil loss due to the better soil health (Reganold et al., 1987; Reganold et al. 2001; Mader et al. 2002; Pimentel 2005). Professor Reganold stated:

"We compare the long-term effects (since 1948) of organic and conventional farming on selected properties of the same soil. The organically-farmed soil had significantly higher organic matter content, thicker topsoil depth, higher polysaccharide content, lower modulus of rupture and less soil erosion than the conventionally-farmed soil. This study indicates that, in the long term, the organic farming system was more effective than the conventional farming system in reducing soil erosion and, therefore, in maintaining soil productivity."

Critics of organic systems point to conventional, no-till production systems as superior to organic systems because the organic systems use tillage. There is only one published study comparing conventional no-till with organic tillage systems. The researchers found that the organic system still had the better soil quality:

"[T]he OR [organic] system improved soil productivity significantly as measured by corn yields in the uniformity trial ... These higher levels of soil C and N were achieved despite the use of tillage (chisel plow and disk) for incorporating manure and of cultivation (low-residue sweep cultivator) for weed control....Our results suggest that systems that incorporate high amounts of organic inputs from manure and cover crops can improve soils more than conventional no-tillage systems despite reliance on a minimum level of tillage." (Teasdale et al. 2007)

The latest improvements in organic low/no till systems developed by the Rodale Institute shows that these systems can deliver high yields as well as excellent environmental outcomes (Rodale 2006).

Organic Systems Use Water More Efficiently

The Millennium Ecosystem Assessment states: "The amount of water impounded behind dams quadrupled since 1960, and three to six times as much water is held in reservoirs as in natural rivers. Water withdrawals from rivers and lakes doubled since 1960; most water use (70% worldwide) is for agriculture." (MA Report 2005)

The current policies of diverting more of the water used for irrigation in Australia back into the rivers for environmental flows, especially in the Murray Darling basin, has major implications for agriculture.

The science around climate change is showing that Australia will see an increase in the frequency and longevity of droughts. The nature of rainfall is predicted to change in many areas. While the amount of rainfall may change slightly, there will be an increase in rainfall patterns causing less frequent rain events, which are heavier in intensity and shorter in duration. This will mean that farming systems will need to be able to capture more of these shorter events and store them in the soil for longer periods. Research shows that organic systems use water more efficiently due to better soil structure and higher levels of organic matter, especially humus (Lotter 2003; Pimentel, 2005):

"Soil water held in the crop root zone was measured and shown to be consistently higher by a statistically significant margin in the organic plots than the conventional plots, due to the higher organic matter content in the organic treated soils.... Data collected over the past 10 years of the FST experiment show that the MNR [organic manure system] and LEG [organic legume system] treatments improve the soils' water-holding capacity, infiltration rate, and water capture efficiency. LEG maize soils averaged a 13% higher water content than CNV [conventional system] soils at the same crop stage, and 7% higher than CNV soils in soybean plots..." (Lotter 2003)

The open structure allows rainwater to quickly penetrate the soil, resulting in less water loss from run off: "The exceptional water capture capability of the organic treatments stood out during the torrential downpours during hurricane Floyd in September of 1999. The organic systems captured about twice as much water as the CNV [conventional] treatment during that two-day event." (Lotter 2003)

Humus can store more than twenty times its weight in water so that rain and irrigation water is not lost through leaching or evaporation (Handrek, 1990; Stevenson, 1998; Handrek and Black, 2002). It is stored in the soil for later use by the plants (Drinkwater, 1998; Zimmer 2000; Mader, 2002). One consistent piece of information coming from many studies is that organic agriculture performs better than conventional agriculture in adverse weather events, such as droughts (Drinkwater, L.E., Wagoner, P. & Sarrantonio, M. 1998; Welsh R., 1999; Lotter, 2003; Pimentel 2005).

It is essential that farming systems produce sufficient yields to ensure that enough food, fiber, and fuel can be produced on existing farmland, so there is no pressure for increased habitat destruction. While some organic systems have lower yields, there are numerous studies showing that best practice organic agriculture can achieve comparable yields—and at times better yields—in comparison to intensive conventional agriculture. (Pretty, 1995; Pretty, 1998a; Welsh, 1999; Reganold, *et al.*, 2001; Parrot, 2002; Leu 2004; Pimentel, 2005; Badgley, 2007; Unep-Unctad, 2008; Bradford, 2008; Posner, 2008).

The assumption that greater inputs of synthetic chemical fertilizers and pesticides are needed to increase food yields is not always accurate. In a study published in *The Living Land: Agriculture, Food and Community Regeneration in the 21st Century*, Jules Pretty Obe looked at projects in seven industrialized countries of Europe and North America:

"Farmers are finding that they can cut their inputs of costly pesticides and fertilizers substantially, varying from 20-80%, and be financially better off. Yields do fall to begin with (by 10-15% typically), but there is compelling evidence that they soon rise and go on increasing. In the USA, for example, the top quarter sustainable agriculture farmers now have higher yields than conventional farmers, as well as a much lower negative impact on the environment." (Pretty 1998)

The following are are three studies that show high yields and beneficial environmental outcomes from organic systems:

1. United Nations Study Organic Agriculture Increased Yields by 116%

The report by the United National Conference on Trade and Development (UNCTAD) and the United Nations Environment Programme (UNEP) found that organic agriculture increases yields in Africa by an average of 116% in total and 128% in East Africa. The report notes that despite the introduction of conventional agriculture in Africa, food production per person is 10% lower now compared to the 1960s: "The evidence presented in this study supports the argument that organic agriculture can be more conducive to food security in Africa than most conventional production systems, and that it is more likely to be sustainable in the long term." (UNEP-UNCTAD 2008)

2. US Agricultural Research Service (ARS) Pecan Trial

Pecans managed organically by the ARS outyielded the conventionally managed, chemically fertilized Gebert orchard in each of the five recorded years. The yields on the ARS organic test site surpassed the Gebert commercial orchard by 18 pounds of pecan nuts per tree in 2005 and by 12 pounds per tree in 2007 (Bradford 2008).

3. Rodale Organic Low/No-Till

The Rodale Institute has been conducting trials on a range of organic low tillage and no-tillage systems. The 2006 trails resulted in organic yields of 160 bu/ac (bushels an acre) compared to the Country average of 130 bu/ac. In an article published in *The Guardian*, Professor George Monbiot reported that in the United Kingdom trials, wheat grown with manure has produced consistently higher yields for the past 150 years than wheat grown with chemical nutrients (Monbiot 2000). The study into apple production conducted by Washington State University compared the economic and environmental sustainability of conventional, organic, and integrated growing systems in apple production and found similar yields: "Here we report the sustainability of organic, conventional and integrated apple production systems in Washington State from 1994 to 1999. All three systems gave similar apple yields." (Reganold *et al.*, 2001)

In an article published in the peer-review scientific journal, *Nature*, Laurie Drinkwater and colleagues from the Rodale Institute showed that organic farming had better environmental outcomes as well as similar yields of both products and profits when compared to conventional, intensive agriculture (Drinkwater 1998). Dr. Rick Welsh, of the Henry A Wallace Institute reviewed numerous academic publications comparing organic production with conventional production systems in the USA. The data showed that the organic systems were more profitable. This profit was not always due to premiums but due to lower production and input costs as well as more consistent yields. Dr. Welsh's study also showed that organic agriculture produced better yields than conventional agriculture in adverse weather events, such as droughts or higher than average rainfall (Welsh 1999). The editorial of *New Scientist* February 3, 2001 stated that low-tech sustainable agriculture is increasing crop yields on poor farms across the world, often by 70% or more. This has been achieved by replacing synthetic chemicals in favor of natural pest control and natural fertilizers (New Scientist, 2001).

Professor Jules Pretty, the Director of the Centre for Environment and Society at the University of Essex in the UK wrote: "Recent evidence from 20 countries has found more than two million families farming sustainably on more than four to five million hectares. This is no longer marginal. It cannot be ignored. What is remarkable is not so much the numbers, but that most of this has happened in the past five to 10 years. Moreover, many of the improvements are occurring in remote and resource-poor areas that had been assumed to be incapable of producing food surpluses" (Pretty, 1998b). Professor Pretty gives other examples from around the world of increases in yield when farmers have replaced synthetic chemicals and shifted to sustainable organic methods.

- 223,000 farmers in southern Brazil using green manures and cover crops of legumes and livestock integration have doubled yields of maize and wheat to 4–5 tons/ha.
- 45,000 farmers in Guatemala and Honduras used regenerative technologies to triple maize yields to 2–2.5 tons/ha and diversify their upland farms, which has led to local economic growth that has in turn encouraged re-migration back from the cities.
- 200,000 farmers across Kenya partaking in sustainable agriculture programs have more than doubled their maize yields to about 2.5 to 3.3 t/ha

and substantially improved vegetable production through the dry seasons.

+ 100,000 small coffee farmers in Mexico have adopted fully organic production methods, and increased yields by half.

+ A million wetland rice farmers in Bangladesh, China, India, Indonesia, Malaysia, Philippines, Sri Lanka, Thailand, and Vietnam have shifted to sustainable agriculture, where group-based farmer-field schools have enabled farmers to learn alternatives to pesticides increased their yields by about 10% (Pretty, 1995).

In the report, *The Real Green Revolution*, Nicolas Parrott of Cardiff University, UK, gives case studies that confirm the success of organic and agroecological farming techniques in the developing world (Parrott, 2002).

+ In Madhya Pradesh, India, average cotton yields on farms participating in the Maikaal Bio-Cotton Project are 20% higher than on neighboring conventional farms.

+ In Madagascar, SRI (System of Rice Intensification) has increased yields from the usual 2–3 tons per hectare to yields of 6, 8, or 10 tons per hectare.

+ In Tigray, Ethiopia, a move away from intensive agrochemical usage in favor of composting has seen an increase in yields and in the range of crops it is possible to grow.

+ In the highlands of Bolivia, the use of bone meal and phosphate rock and intercropping with nitrogen-fixing Lupin species have significantly contributed to increases in potato yields.

Climate Change

Climate change is emerging along with habitat loss as one of the priority environmental issues. The fact is that the latest studies clearly show that the world is warming and that the rate of greenhouse gas emissions continues to increase despite international efforts like the Paris Agreement. Greenhouse gases reached a new record of 400 ppm in 2016, their highest in 800,000 years. They have been increasing by 2 ppm per year, and in 2017, they increased by a record 3.3 ppm.

Even if the world stopped polluting the planet with GHGs it will take many decades to reverse climate change. This means that farmers have to adapt to the increasing intensity and frequency of adverse and extreme weather events such as drought and rainfall. Many areas of the planet are experiencing exactly this. Published studies show that organic farming systems are more resilient to the predicted weather extremes. The studies showed that organic systems can have higher yields than conventional farming systems in weather extremes (Drinkwater, 1998; Welsh R., 1999; Pimentel D., 2005). The Wisconsin Integrated Cropping Systems Trials found that organic yields were higher in drought years and the same as conventional in normal weather years (Posner *et al.*, 2008). The Rodale Farm Systems Trial (FST) showed that the organic systems produced more corn than the conventional system in drought years:

> "Average corn yields in those 5 dry years were significantly higher (28% to 34%) in the two organic systems: 6938 and 7235 kg per ha in the organic animal and the organic legume systems, respectively, compared with 5333 kg per ha in the conventional system." (Pimentel D., 2005)

The researchers attributed the higher yields in dry years to the ability of soils on organic farms to better absorb rainfall. This is due to the higher levels of organic carbon, making the soils more friable and better able to store and capture rain: "This yield advantage in drought years is due to the fact that soils higher in carbon can capture more water and keep it available to crop plants." (La Salle and Hepperly, 2008)

Energy Use Efficiency and Conservation

Two published studies (Mader *et al.*, 2002; Pimentel 2005), in peer reviewed scientific journals, of long-term comparison trials (21 and 22 years) of conventional and organic systems found that the organic systems use less fossil fuels and, therefore, emit significantly lower levels (around 30%) of greenhouse gases. The long-term apple comparison trial conducted by Reganold, in Washington, USA, showed that the organic system was more efficient in its energy use.

> "When compared with the conventional and integrated systems, the organic system produced sweeter and less tart apples, higher profitability and greater energy efficiency" (Reganold et al., 2001).

Rodale Institute's organic rotational no-till system can reduce the fossil fuel needed to produce each no-till crop in the rotation by up to 75% compared to standard-tilled organic crops. (LaSalle, T. and Hepperly, P. 2008)

Table 1: Energy Used in Different Corn Production Systems in Liters of Diesel Per Hectare (l/ha)

Conventional Tillage:	231
Conventional No-Till:	199
Organic Tillage:	121
Organic No-Till:	77

The critical information here is that the system with the lowest energy use also has the highest yields. This shows the potential for these new low input organic systems to ensure that the world can adapt to climate change and produce sufficient food to feed all living beings. This very important information should be picked up by governments and research institutions around the world to ensure that this can improved and adapted to all agricultural regions on the planet (Pimentel *et al.*, 2005).

Soils as a Carbon Sink

Soils are the greatest carbon sink after the oceans. According to Professor Rattan Lal of Ohio State University, "although the figure is frequently being revised upwards with new discoveries, over 2,700 Gt of carbon is stored in soils worldwide, which is well above the combined total of atmosphere (780 Gt) or biomass (575 Gt), most of which is wood." (Lal 2008)

The amount of CO_2 in the oceans is already causing a range of problems, particularly for species with calcium exoskeletons such as coral. Scientists are concerned that the increase in acidity caused by higher levels of CO_2 is damaging these species and are therefore concerned about the future of marine ecosystems such as the Great Barrier Reef. The world's oceans, like the atmosphere, cannot absorb more CO_2 without causing serious environmental damage to many aquatic ecosystems (Hoegh-Guldberg *et al.*, 2007). Studies show that the widespread adoption of regenerative organic farming systems can reverse climate change. Section 4 "Climate Change Solutions" explains how to do this.

One of the major debates around soil carbon is based on how it can meet the Kyoto Protocol Clean Development Mechanism (CDM) 100-year permanence requirements. Soil carbon is complex mix of fractions of various carbon compounds. Two of these, humus and charcoal (char), are very stable, with Australian research showing that they can last for thousands of years in the soil (Handrek 1990). Other fractions are less stable (labile) and can be easily volatilized into CO_2.

Soil carbon tends to volatilize into CO_2 in most conventional farming systems. However, the correct management systems can continuously increase both the stable and labile fractions. This is due to numerous reasons with several of them discussed in later sections. The research conducted by Dr. Christine Jones at Winona showed that the majority of the newly increased soil carbon was in the stable fractions, "78% of the newly sequestered carbon is in the non-labile (humic) fraction of the soil - rendering it highly stable." (Jones, 2011)

Long-term research conducted for more than 100 years at the Rothamsted Research Station in the UK and the University of Illinios Morrow Plots in the USA showed that the total soil carbon levels can steadily increase and then reach a new stable equilibrium in farming systems that use organic matter inputs (Handrek, 1990). This means that good organic management systems can increase and maintain the labile fractions as well as the stable fractions over the time periods demanded by the CDM.

Chemical agriculture treats soil as inert and an empty container for chemical fertilizers. The new paradigm recognizes soil as living, in which billions of soil organisms create soil fertility. According to Dr. Elaine Ingham, just one teaspoon of compost-rich organic soil may host as many as 600 million to 1 billion helpful bacteria from 15,000 species. Ingham notes that on the flip side, one teaspoon of soil treated with chemicals may carry as few as 100 helpful bacteria.

Chemical agriculture destroys biodiversity; ecological agriculture conserves and rejuvenates biodiversity and is based on the multiple ecological functions that biodiversity performs. Chemical agriculture depletes and pollutes water; organic farming conserves water by increasing the water-holding capacity of soils through recycling organic matter.

Biodiversity and soils rich in organic matter is the best strategy for climate mitigation, climate resilience, and climate adaptation, as shown in *Soil Not Oil*. While lowering the ecological footprint, regenerative organic agriculture increases output when measured through diversity and multifunctional benefits instead of the reductionist category of "yield."

1.5 *The Environmental Benefits of Organic Farming*

Studies show that by not using soluble fertilizers and pesticides, and instead focusing on building high soil humus content and implementing soil conservation techniques, that there is minimal soil and nutrient run-off as well as higher biodiversity on organic farms (Reganold *et al.*, 1987; Zimmer 2001; Reganold *et al.*, 2001; Drinkwater, 1998; Welsh 1999).

A 21-year long comparison study by Swiss researchers (Mader *et al.*, 2002) published in *Science* showed that organic farming is more energy-efficient than conventional farming. The study found that organic fields have healthier soil and greater diversity and number of organisms, including earthworms, beneficial fungi, beetles, and wild plants. A long-term study conducted by the Washington State University published in the science journal *Nature* showed that the negative environmental impact of conventional farming systems was 6.2 times higher than organic systems (Reganold *et al.*, 2001). A viable income is an essential part of the environmental sustainability of farming systems.

The IAASTD reports showed that failure of markets to value environmental services was one of the major reasons for the environmental damage and recommended that governments need to take a range of actions to reverse this ever-increasing degradation. These include a mix of regulatory and market-based solutions:

> "Agriculture generates large environmental externalities, many of which derive from the failure of markets to value environmental and social harm and provide incentives for sustainability. AKST [agricultural knowledge, science, and technology] has great potential to reverse this trend. Market and trade policies to facilitate the contribution of AKST to reducing the environmental footprint of agriculture include removing resource use–distorting subsidies; taxing externalities; better definitions of property rights; and developing rewards and markets for agro-environmental services, including the extension of carbon financing, to provide incentives for sustainable agriculture."
> (IAASTD, 2008)

Organic farming clearly fits well within the IAASTD definition of AKST (agricultural knowledge, science, and technology) and has great potential to reverse the trend towards ever-increasing environmental degradation. Commercial market-based drivers are always the most effective as they are not reliant on changes in government policy and the competing funding complexities of politicians and bureaucrats.

Profitability gives farmers the incentive to continue with good practices and remain sustainable over the long term. Published studies comparing the income of organic farms with conventional farms have found that the net incomes are similar, with best practice organic systems having higher net incomes (Cacek 1986, Wynen 2006). The US Department of Agriculture's National Agricultural Statistics Service conducted a survey of 14,540 US organic farms and ranches in 2008. Organic operations had an average of $217,675 in sales, compared to $134,807 for all US farms as reported in the 2007 Census of Agriculture (Organic Production Survey, 2008).

A study in the US by Dr. Rick Welsh of the Wallace Institute has shown that organic farms can be more profitable. The premium paid for organic produce is not always a factor in this extra profitability. This profit was not always due to premiums but due to lower production and input costs as well as more consistent yields. Dr. Welsh analyzed a diverse set of academic studies comparing organic and conventional cropping systems. Among the data reviewed were six university studies that compared organic and conventional systems (Welsh 1999). The study into apple production conducted by Washington State University showed that the break-even point was nine years after planting for the organic system and 1–16 years respectively for conventional and integrated farming systems: "When compared with the conventional and integrated systems, the organic system produced sweeter and less tart apples, higher profitability and greater energy efficiency." (Reganold et al., 2001)

Research conducted in Australia by agricultural economist Dr. Els Wynen has found similar results in studies conducted in Europe, the US, and Africa, showing higher levels of financial returns for organic production systems (Wynen 2006). A United Nations report found that "Organic production allows access to markets and food for farmers, enabling them to obtain premium prices for their produce (export and domestic) and to use the additional incomes earned to buy extra foodstuffs, education and/or health care." The report noted: "A transition to integrated organic agriculture, delivering greater benefits at the scale occurring in these projects, has been shown to increase access to food in a variety of ways: by increasing yields, increasing total on-farm productivity, enabling farmers to use their higher earnings from export to buy food, and, as a result of higher on-farm yields, enabling the wider community to buy organic food at local markets." (UNEP-UNCTAD 2008)

Conclusion

A large body of published science shows that good practice organic agricultural systems are amongst the most environmentally sustainable of our current agricultural systems. They have the highest biodiversity, do not use the environmentally problematic inputs of pesticides and soluble fertilizers, have the least runoff and soil loss, reduce greenhouse gases, are superior at capturing and storing rainfall, and achieve good yields of high-quality produce that can be sold for viable prices.

Conversion to organic farming is an effective solution to many environmental problems caused by some of our current farming systems. The United Nations, Food and Agriculture Organization stated, "Organic agriculture should be considered simply as the most appropriate starting point… Its widespread expansion would be a cost-efficient policy option for biodiversity" (FAO 2003). The long-term comparison study by Washington State University concluded: "Our data indicate that the organic system ranked first in environmental and economic sustainability, the integrated system second and the conventional system last." (Reganold *et al.*, 2001)

Research from around the world demonstrates that best practice organic agriculture can be high yielding, especially in climatic extremes such as droughts and floods that are predicted to increase in frequency and severity due to climate change.

2

SECTION   **Seeds of Biodiversity**

2.1 *Seeds: The Source of Life in Abundance and Renewal*

Seed is known as *bija* in Sanskrit and Hindi, *shido* in Japanese, *zhangzi* in Chinese, *semi* in Italian, *semilla* in Spanish, *semence* in French, and *der saat* in German. Seeds are the source of life and the first link in the food chain; they are the basis of all life forms in the universe. Seeds embody millennia of evolution, thousands of years of farmers' breeding, and the culture of freely saving and sharing seed. It is the expression of Earth's intelligence and the intelligence of farming communities down the ages. A seed renews itself over time as it grows into a crop from which comes a new seed.

A seed is a living organism, even though it looks inert. To remain alive, the embryo must have access to food and oxygen. If it runs out of food or is subjected to physical damage, including an attack by insects or fungi, it will die. The life and vigor of seeds can be shortened or extended depending on how they are treated. They can be physically damaged before or during harvest, transport, or storage, and their longevity can be dramatically shortened if they are not stored in good conditions.

Most of the world's agricultural diversity took humankind over 10,000 years to create, but we may lose most of it in a single generation. Until the early 20th century, food for human beings was provided by as many as 10,000 different species of plants, each further represented by thousands of different cultivated varieties. But today, over 90% of the world's nutrition is provided by just 30 different plants, whereas 75% of the total calories consumed by humankind are provided by only four crops: wheat, rice, corn, and soybean. In the past, diverse strains that strengthened each local ecosystem have been replaced by only a handful of super-hybrid Green Revolution varieties grown through worldwide monocropping.

Seeds embody the ideas, knowledge, culture, philosophy, tradition, and heritage of a people. Seeds thus represent the wisdom of the years of farmer's research who have meticulously worked in perfect coordination with nature, considering the climate and hydro-geological parameters of the region. There exists a complete harmony in the ecological niche of the crop grown in the region.

Biodiversity of seeds, therefore, goes hand in hand with the biodiversity of knowledge systems and the biodiversity of paradigms of breeding.

For over 10,000 years, Indian farmers have used their brilliance and Indigenous knowledge to domesticate and evolve thousands of crops, including 200,000 rice varieties, 1,500 wheat varieties, 1,500 banana varieties, and hundreds of species of dals, oilseeds, mangos, millets, pseudocereals, vegetables, and spices.

This brilliance in breeding was abruptly stopped when the chemical industry imposed the Green Revolution on us in the 1960s with its roots in the war. As in the colonization of the past, our intelligence in seed breeding and agriculture was denied; our seeds were called "primitive" and displaced. A mechanical "intelligence" of industrial breeding for uniformity and external inputs was imposed. Instead of evolving diverse varieties of diverse species, our agriculture and diet were reduced to rice and wheat.

Our native seeds have been bred for resilience. Industrial seeds are bred for chemical monocultures. They are water-intensive and vulnerable to failure in times of drought. To deal with climate change and water scarcity, we need to cultivate varieties of millet, which use less water, and increase the soil's water holding capacity while increasing our food and nutrition security.

Scientifically, a seed is a small embryonic plant enclosed in a covering called the seed coat, usually with some stored food. Seeds are also referred to as germplasm. Germplasm is the set of varieties, both modern and traditional (Indigenous or landraces) of a crop species, including their wild progenitors—when still available.

Seed varieties are subgroups within species, either the products of natural selection or selections made by humans for particular traits. The varieties are usually distinguished in "modern" or "improved" varieties (i.e., selected during the breeding programs conducted by the researchers) and ancient and Indigenous varieties (i.e., selected and, in many cases, preserved by farmers). This is, of course, a misnomer since farmers' varieties that have evolved and are still used today are modern varieties also because they are being used in modern times. A difference usually observed between ancient, traditional varieties and "modern" varieties is genetic uniformity. "Modern" varieties are bred for uniformity, while farmers have bred their varieties for diversity. Older varieties are also better adapted to more marginal conditions, such as less fertile soils or drought. Diversity and evolutionary potential create resilience to pests, diseases, and climate change, while uniformity contributes to and is vulnerable to these factors.

2.2 *Farmers: The First Link in Plant Breeding*

Seeds of agricultural crops have been developed over centuries by farming communities across the world. These seeds have been freely exchanged with other communities around the globe and have led to the development of new varieties. Today, with the entry of the corporate sector in seed production, supply, and new seed production technologies, seed varieties have been given a variety of names depending on who evolved it, how it was evolved, and its potential for making profits.

For millennia, farmers have studied, identified, modified, cultivated, and exchanged seeds freely so that they may provide the best food for nutrition and taste. In this capacity, the farmer has always been a scientific plant breeder. Farmers have traditionally conserved and developed diversity in their fields through the ongoing cultivation of the varieties. Through plant domestication, by performing the role of the plant breeder, farmers have created our food and fiber crops and developed thousands of crop varieties. Plant breeding has become professionalized in the current era, and most farmers rely on seed companies for their annual seed needs.

While farmers breed for diversity, corporations breed for uniformity; while farmers breed for resilience, corporations breed vulnerability. While farmers breed for taste, quality, and nutrition, industry breeds industrial processing and long-distance transport in a globalized food system.

Industrial breeding has used different technological tools to consolidate control over the seed, from so-called High Yielding Varieties (HYVs) to hybrids, genetically engineered seeds, "terminator seeds," and now, synthetic biology. The tools might change, but the quest to control life and society does not.

As the farmer produced mainly for the family, village, and the rest of the larger community—with the central vision being sustainability of both lifestyle and nature, including land and water resources—they were interested in conserving the plant varieties they developed.

2.3 *Ex Situ and In Situ Methods*

Today there are three sources of seed supply:

1. The farmer has historically been the producer of perennial varieties, which could reproduce themselves perpetually
2. Public sector research institutions have bred short-term varieties for "high yield." These seeds could be saved and used by the farmer for some time, but their yield reduces after a few years
3. Transnational corporations produce non-renewable and, therefore, non-sustainable seeds through hybrids and tissue culture, or GMOs where the farmer has to return to the company for fresh seed each time they have to sow. The seed corporations are also chemical corporations that have merged into three dominant groups: Bayer Monsanto, Dow Dupont, and Syngenta ChemChina

The two main reasons farmers have lost control over diversity, seeds, and agriculture are:

1. *Viewing agriculture not as a sustainable lifestyle but as the production of commodities for the market economy*: When only the marketable product of farming became the focus of agriculture, the paramount need was to produce more and more grain. This led to monoculture and the loss of many varieties that the farmer needed outside input to ensure the fertility of their soil, protection for their crops, and subsistence.
2. *The development of seeds shifted from the farmers to the scientists*: With high yield as the main focus, seeds were developed to produce more grain at the cost of the other valuable parts of the plant, such as leaves and straw. As this grain could be produced only in the presence of intensive use of chemical fertilizers, the farmer became dependent upon the government and the fertilizer industry. These inputs cost money, leading the farmer to depend upon subsidies and bank credits.

The state also took over the marketing of the products, and the direct link between the consumer and the producer was broken. The state purchased the products from the producer and then passed them onto the consumer. As the remuneration for the products was not calculated to include all the farmer's costs, the farmer was never in a position to pay off their debts and regain their independence. Farming has thus become economically not viable. Thus, the farmer lost control over the diversity within their field, as well as over what to produce, and when, how, and for whom to produce it.

The Politics of Language

As long as the farmer was recognized as both the chief plant breeder and the primary source of seed supply, the seed was viewed as a commons that the farming community retained collective ownership over and, therefore, control over biodiversity. When the corporate sector moved into the seed sector to create markets by replacing both farmers' renewable varieties, a new phraseology for seed was created to justify the separation of the farmer and farmers knowledge from the seeds; thus erasing the contribution farmers have made to breeding, and appropriate farmers' varieties as "raw material" for industrial breeding by corporations. This mindset is revealed through the use of phrases like "landraces," "germplasm," and the definitions of "variety," "high yield," "innovation," and "intellectual property."

Farmers' varieties are those varieties which have been developed by farmers and grown for many generations to suit their ecological, nutritional, medicinal, and resource needs. Their physical and genetic qualities are relatively stable. These have sometimes been called "landraces" to distance them from the contributions that farmers have made towards their evolution through selection.

This term suggests that such varieties have popped up from the land, like wild species, and farmers have made no intellectual and creative contributions to breeding. They have also derogatorily been called "primitive cultivars" in contrast to "elite cultivars," evolved by scientists. Farmers' varieties, like any other seed variety, are an embodiment of intellectual contribution. Farmers' varieties are perennial and sustainable. They are also referred to as Indigenous, native, heirloom or heritage, open-pollinated, and in India, *jwaari*, *nate*, or *desi* seeds.

Erasing farmers' contributions as breeders also erases the community's rights to seed. It transforms seeds from community-managed commons with clear rules of conservation and sharing into an open access system or the "common heritage of humankind," which can be freely grabbed by the powerful corporations and made into their intellectual property. This creates seed monopolies and biopiracy, causing seed wars in the FAO in the 1980s and 1990s.

The fact is that farmers are the first breeders; many varieties have evolved because of their intervention over centuries. "Landraces" is a scientifically and epistemically inaccurate term. A more accurate term is "farmers' varieties." The use of the term "farmers' varieties" recognizes the plant breeder role of the farmer and removes such varieties from the realm of "common heritage of mankind" to the commons conserved, managed, used sustainably, and owned collectively by farming communities. These commons must be protected against enclosures and piracy.

While the use of the term "landraces" justifies intervention by corporations, the use of the term "farmers' varieties" makes it clear that the innovators are the farmers. If rewards are to be given for innovation, they should be given to the farmers.

Heirloom seeds

"Heirloom" refers to a variety of plant or animal that has been passed down from generation to generation. Usually, a minimum of three (human) generations are required for a plant to be known as an heirloom, but the term may also refer to old (more than 100 years) commercial varieties. All heirlooms are open-pollinated, but not all open-pollinated varieties are heirlooms.

Variety

The word "variety" in today's agriculture does not refer anymore to the vast diversity that exists today due to farmers' innovations over centuries. It has come to represent standardization, which needs protection by legislation like those created by the International Union for the Protection of New Varieties of Plants (UPOV). In fact, in order to be eligible for protection, under plant variety protection legislation, a variety must be:

+ **New:** The variety must not have been exploited commercially. Novelty is defined commercially, not in evolutionary terms
+ **Distinct:** It must be clearly distinguishable from all other varieties known at the date of application for protection
+ **Uniform:** All plants of the variety must be sufficiently uniform to allow them to be distinguished from other varieties taking into account the method of reproduction of the species
+ **Stable:** It must be possible for the variety to be reproduced, unchanged

By its very nature, this definition rules out farmers' varieties and destroys biodiversity by producing uniformity as a necessity.

The criteria for industrial breeding and industrial agriculture are called Distinctiveness, Uniformity, and Stability (DUS). It is based on the intensive use of chemicals, water, and fossil fuels. DUS ignores the need for diversity, nutrition, safety, and the need to create low-cost sustainable livelihoods in the context of economic collapse and slowdown and the consequent need to localize food systems. When India became a member of WTO, we ensured that we did not adopt the UPOV but evolved a *sui generis* system called The Plant Variety Protection and Farmers Rights Act in 2001.

Germplasm Sarkari

This term is used to denote the genetic material within the plant, particularly in the seed. Such seeds are also derogatorily called "primitive seeds." The term "germplasm" is both scientifically inaccurate and reinforces the negation of farmers' rights by devaluing farmers' varieties. It was created by the German biologist August Weismann who used it to refer to hereditary components of organisms which he assumed to be totally separate from the body of the organism as well as from the environment.

The assumption of the genetic material of "germplasm" as insulated totally from the organism and the environment has been proven scientifically to be false. Genetic traits and genes are not independent entities, but dependent parts of the whole organism that gives them effect, plants in this case an organism, in turn, interacts with and evolves in an environment which influences it. This influence is, in fact, the source of biological diversity. It is now also recognized in the emerging science of epigenetics.

The term "germplasm" continues to be used today to further distance the farmer from the source of genetic material: seeds of diverse varieties. It assumes that while the seed is itself worthless to the farmer because it is not "improved" or "high yield," it contains "germplasm," or rather genetic material, that scientists and corporations can use to improve plant varieties for the benefit of the farmer. Conservation of this plant becomes the concern of international and national gene banks, rather than concern for farmers, who for centuries have been conserving this diversity in the farm itself.

Besides negating the plant breeding role of the farmer, such distancing of the farmer from variety turns them into a consumer of seeds produced by corporations as the only varieties that become available to them are developed by corporations, mostly seeds of plants that cannot reproduce themselves.

High Yielding Varieties (HYVs)

These varieties for Green Revolution seeds is a misnomer because the term implies that the seeds are high yielding in and of themselves. The distinguishing feature of these seeds, however, is that they are highly responsive to certain key inputs such as fertilizer and irrigation. They are actually the high response varieties. Though these seeds can be saved by farmers, they are non-sustainable due to vulnerability to diseases and pests and therefore need to be replaced after every few years. These seeds are also called "*sarkari*," "*vikas*," "society," or even "government" seeds as they have been developed are distributed primarily by the public sector.

Hybrid seeds

These seeds are the first-generation seeds (Fl) produced from crossing two genetically different parent species. The progeny of these seeds cannot economically be saved and replanted as the next generation does not breed true and will give much lower yields.

Hybridization is only one of the breeding techniques. It does provide high-yielding varieties, but so do other breeding techniques. Why did hybridization gain such predominance over other methods? Using the example of hybridization of corn in the US, Jack Kloppenburg in *First the Seed* explains:

> "[T]here is an even more compelling reason to examine closely the historical choice of breeding methods in corn, for the use of hybridization galvanized radical changes in the political economy of plant breeding and seed production. There is a crucial difference between open-pollinated and hybrid corn varieties: Seed from a crop of the latter, when saved and replanted, exhibits a considerable reduction in yield. Hybridization thus uncouples seed as "seed" from seed as "grain" and thereby facilitates the transformation of seed from a use-value to an exchange value. The farmer choosing hybrid varieties must purchase a fresh supply of seed each year."

Hybridization is thus like biologically patenting the seed. No one else, neither the farmer nor a rival company, can produce exactly similar seeds unless they know the parent lines, which are the company's secrets. This characteristic of the hybrid seed has been fundamental to the rapid growth of the American seed industry. The corporate seed sector in India is also involved, mainly in developing hybrid seeds, including seeds of corn, sorghum, vegetables, and food grains. The last is called the biological patenting of seed. Patents give the seed owner the exclusive right to multiply, save, develop different varieties, and sell seeds. Biological patenting effectively prevents the farmer from multiplying, preserving, and selling the seed.

When hybrid seeds are still being developed, the farmer still has some control over the seed. Agribusiness uses legal patents in agriculture to take over this control.

New Biotechnologies and Genetic Engineering Technologies

The new biotechnologies include the tissue or cell culture, cloning and fermentation methods, cell fusion, embryo transfer, and recombinant DNA technology (genetic engineering).

Tissue or cell culture

This is among the most commonly used new technologies. Tiny pieces of plant material—tissue or isolated cells—are grown in an artificial medium that keeps them alive. Particular hormones like rooting hormones help them to develop into complete plants. These baby plants are identical to the parent plant and each other.

Cloning and fermentation

Cloning is the process of forming a cell culture starting from a single cell that can multiply itself. The culture thus contains cells with identical characteristics. Each of these cells can then be used to propagate new plants through tissue culture. Fermentation generally means a natural process in which the biological activity of a microorganism (bacteria or virus) is vital, for example, making yogurt, wine, or other products. Such processes using genetically engineered bacteria can produce vanilla, jasmine, and citrus fragrances out of a totally unrelated medium. Using this method, edible oil can be converted into cocoa butter.

Cell fusion and embryo transfer

These technologies are used mainly for dairy and livestock breeding purposes.

Recombinant DNA technology (genetic engineering)

This technology involves transferring genes from one cell to another. Genetic engineering crosses the boundaries of nature by allowing genes from one life form to be introduced into an unconnected life form, e.g., genes from fireflies have been introduced into tobacco to create a variety that glows naturally; genes from a fish found in the Arctic Ocean have been introduced into soybeans and tomatoes so that those plants can withstand cold and frost and also be refrigerated for long periods. Genes have also been introduced into plant varieties to make them resistant to a particular brand of herbicide.

Genetically engineered cells are mass propagated through tissue culture methods to produce thousands of new life forms with unique characteristics. Such life forms are often called transgenic.

The new biotechnologies are even more disruptive of the social fabric as they further distance the farmer from seed development. Any development takes place not merely in laboratories but within the seed itself. The farmer becomes further dependent on outside agents for resources and information about how to them.

The seeds produced by the new technologies are in no way superior to either farmers' varieties or the seeds of the Green Revolution. By their very nature, they are monocultures and will therefore have the same vulnerability to diseases and pests.

As their characteristics have been modified at the level of the gene, their progeny will have the same characteristics. Thus, a plant that is engineered to produce its pesticide will pass on this property to its progeny, who will continue to release it into the environment irrespective of any harm that it can cause. Further, products of genetic engineering have not been tested for adequate periods to see their long-term effects.

Genes are a segment of DNA carrying very specific information about plants. The number of genes may vary from a few dozen genes (virus) to tens of thousands (higher plants and animals). Some genes carry information for activating other genes. Of the few 100,000 genes of the tomato plant, which is one of the most studied, only around 300 genes have been identified so far. DNA is a molecule that is supposed to carry all the information necessary to create a new identical plant. The new biotechnologies assume that the DNA alone makes new life and itself is not influenced by any external environment. However, it has been shown that DNA cannot reproduce itself without the intervention of other necessary information and enzymes. It is thus influenced by the environment and continues to evolve by existing in the environment.

Types of Pollination

Living organisms are complex self-organizing systems. Whether genes are added, edited, or removed through genetic engineering, it disrupts the self-organizing capacity of living systems.

A few years ago, a new genetic engineering tool called CRISPR was developed through Gates funding. CRISPR is short for CRISPR/cas9, which is short for Clustered Regularly-Interspaced Short Palindromic Repeats/CRISPR associated protein 9. It is a combination of a guide RNA and a protein that can cut DNA.

This paradigm is the old genetic reductionism. The aim is to rush to patents and ownership and to escape regulation. CRISPR has been described as "a relatively easy way to alter any organism's DNA, just as a computer user can edit a word in a document."

However, a seed and a living organism are not computers. People write and edit documents with the facility of computers, but the complex self-organization of living systems is written by the living system.

Open-pollinated seeds

Open-pollinated plants are allowed to reproduce according to the impulse of pollinators like bees, the wind, or other pollination mechanisms they depend upon. "Open-pollinated" can also refer to self-pollinating plants (e.g., rice, tomatoes, lady's finger, and beans) or cross-pollinating plants (corn, papaya, brinjal, gourds, cabbages, carrot, etc.). This term is usually used to describe plants that are not hybrids. Open-pollinated seeds can be just as vigorous, disease-resistant, and commercially useful as hybrids if properly saved.

Self-pollinated plants

Self-pollinated plants pollinate themselves to make fertile seeds. These plants have both male and female parts in the same flower. Often, self-pollinating plants pollinate themselves even before the flower has opened. Many vegetables and small grains are self-pollinated, which means that these species will "automatically" pollinate themselves into thousands of pure breeding lines (genetically uniform lines) after the initial cross or mutation has occurred. Self-pollinated stock can either be a pure line or a multi-line.

A pure-line variety is totally uniform, the progeny of one initial plant, which is then multiplied out to thousands of seeds. These populations are almost completely uniform and are often offered by significant commercial seed sources. Selection in this line is ineffective because there is not enough genetic diversity from which to select. These populations change slowly over time, and after dozens of generations, they might show slight variations due to mutations or chance crosses.

A multi-line population is a collection of pure-line plants. When working with these types of populations (often seen in heirloom varieties), one can effectively select between different lines and make progress. Within these populations, with the low occurrence of cross pollination, there is still enough genetic variation.

In such fields, where there is a diversity of blooms and habitats for native and domesticated pollinators, cross-pollination within the self-pollinated plants is possible. Even though the chances are less, but with certain varieties of peppers and tomatoes, chances of cross-pollination are quite high.

Cross-pollinated plants

Cross-pollinated plants are plants that have flowers designed to promote pollination with other plants. This is an evolutionary mechanism to prevent inbreeding. Inbreeding can cause all sorts of problems, such as loss of vigor, loss in yield, etc. The evolutionary strategy of cross-pollination is to experiment, mingling and mixing with other genotypes to cover up deleterious traits and increase fitness to its environment.

Because these plants thrive on variation, it takes special efforts to inbreed these species into a stable variety. In contrast to self-pollinated crops, cross-pollinated crops have a different mother plant from the father plant. Traits can be elusive in cross-pollinated plants. Every time an individual plant cross-pollinates, its genes recombine, often resulting in unpredictable combinations for the next couple of generations.

Table 2: Pollination behavior of some crops

	FIELD CROPS	VEGETABLE CROPS
Self Pollinated	Paddy	Cowpea
	Wheat	Clusterbean
	Ragi	Tomato
	Barley	Dolichosbean
	Oats	French bean
	Blackgram	Garden pea
	Greengram	Lettuce
	Bengalgram	
	Groundnut	
	Soybean	
	Jute	
Cross Pollinated	Maize	Cabbage
	Bajra	Carrot
	Sunflower	Cucurbits
	Safflower	Onion
	Niger	Radish
	Castor	Amaranthus
	Mesta	
	Sunhemp	
	Mustard	
Often-Cross Pollinated	Sorghum	Bhindi
	Redgram	Brinjal
	Sesamum	Chilli
	Cotton	Capsicum
	Lucerne	Sweet Pepper
	Berseem	Limabean

Why Save Seeds?

The rate of ecological destruction of biodiversity has been recognized today, as has the need for conservation at the farmers and the state level.

The present model of industrial agricultural development has split agriculture into three different kinds of activities engaged in by three different types of institutions and actors:

1. Farmers as consumers of high-cost seed and chemical inputs for industrial agricultural production.
2. Breeders, involved in plant breeding for chemical inputs and breeding uniformity. During the Green Revolution, the breeders were public institutions. With globalization and the introduction of genetic engineering, corporations have become the leading players in the seed sector.
3. Ex situ gene banks, involved in the conservation of genetic resources.

There are two types of conservation activities based on two paradigms of breeding. One type of conservation is when seeds and propagating material of plants are collected by groups of people (not necessarily farmers) and are stored in particular gene banks. Here the farmer is merely the supplier of the genetic material to be kept under high tech conditions for future use by breeders and seed companies. They are totally distanced from their role as a plant breeder. The only other role envisaged for the farmer is that of a consumer of seeds as non-renewable commodities, which in turn is linked to food as a commodity.

The second type of conservation starts and ends in the farmers' fields. Such conservation is carried out within the environment where the diversity grows. The farmer, as a breeder, selects and conserves agricultural diversity by growing it in their fields.

The Navdanya philosophy treats the farmer as a breeder and an expert in their own right. While the dominant system of agriculture does not recognize farmers' contribution to breeding and therefore awards breeders' rights only to the seed industry or to researchers, in the partnership model promoted by Navdanya, farmers and scientists are equal partners, and seed is a commons.

Evolutionary and Participatory Plant Breeding

This method is based on the creation of large mixtures: mixing both crosses or old varieties, or mixing old and new varieties (the composition of the mixture is discussed with farmers), and leaving such mixtures to evolve under conditions in which the farmer wishes to cultivate future varieties in an organic system.

Navdanya recognizes that in situ conservation is a political commitment and cannot be sustained through subsidies and external support alone. However, given the dominant agriculture models which destroyed both diversity and the farmer's role in the production economy, catalytic actions are needed to create awareness of the magnitude and consequences of the erosion of biological diversity, local community rights to biodiversity and cultures, and the knowledge that makes conservation of diversity possible.

For over 30 years, Navdanya has practiced and promoted farmers' selection of local varieties of seeds adapted to the climatic condition of their region. Through natural selection, Navdanya has developed improved varieties of 730 rice varieties, 15 millets, 12 pulses, 27 vegetable varieties, 100 wheat varieties, 15 barley varieties, and 8 oilseeds. Natural selection acts on a population by modifying it gradually and continuously making it better adapted.

Through evolutionary and participatory breeding, Navdanya has selected more than ten new varieties of paddy and two new varieties of barley, namely *pillu motta, bindu dhan, ruby, neeli saathi, barik naj, pink nan, godiyal.* The results of several experiments conducted by Navdanya with rice have shown that some characteristics like size of the grain, length of the straw, the color of grain, maturing time early or late, aromatic properties, resistance to water logging, and drought tolerance increase in these mixtures over time due to natural selection.

Saving Seeds to Preserve Biodiversity

The rate of ecological destruction and the accompanying loss of biodiversity has forced nations and international organizations to make efforts to prevent it.

Article 14.2 of Agenda 21 states that "Major adjustments are needed in agricultural, environmental and macroeconomic policy, at both national and international levels, in both developed as well as development" (Art. 15.2). Recognizing that "Farmers' fields and gardens are also of great importance as repositories" of biodiversity, the agenda calls for both in situ and ex situ conservation, and invests the national governments the "responsibility to conserve their biodiversity and use their biological resources sustainably." (Art. 15.3).

There are two types of conservation activities: One type is farm-based, where the farmer conserves a variety by continuing to cultivate it regularly. This kind of conservation is called in situ conservation.

The second kind of conservation is when seeds and propagating material of plants are collected by groups of people (not necessarily farmers) and stored in special gene banks again away from the field. This kind of conservation is called ex situ conservation.

2.3 Ex Situ and In Situ Methods

The risks of breeding towards uniformity led to the emergence of gene banks. The International Bureau of Plant Genetic Resources (IBPGR) was set up in 1974 under the aegis of the Consultative Group on International Agricultural Research (CGIAR) with a mandate to promote an international network of genetic resource centers to further the collection, conservation, documentation, evaluation and use of plant germplasm. At the national level, the National Bureau of Plant Genetic resources (NBPGR) was set up for collection and ex situ conservation of crop varieties.

> Though Art. 15.1 of the Biodiversity Convention, the highest ranking treaty on biodiversity in the world today, recognises national sovereignty over natural resources, institutions like ICRISAT and gene banks are exempt from national scrutiny. In effect, they are authorised to collect diversity from farmers' fields but are not compelled to return it to them. They can, however, under the phrase "access to genetic resources" give the same away to international agencies and private business, as when Dr. Richaria's indigenous seed collection was given away to the Philippines with the active collaboration of the Ford Foundation and the ICAR.

Gene banks cannot safeguard our diversity because they are centralized and accessible to those who privatize the genetic wealth, and because in times of climate change, seeds need to keep evolving. Plant genetic material is being stolen from gene banks with impunity. In 1970, IRRI's material was stolen and now hybrid rice has been stolen from agricultural universities. The records from 1946 onwards, which list what is missing and where it has gone, list less than 1% of what is missing.

Thus, it is clear that while gene banks and public sector breeders of seed collect biodiversity from farmers' fields, they do not make this available to the farmer. Instead, diversity flows from farmers' fields to gene banks and then onto breeders, but it is systematically eroded at the source. This erosion makes agriculture vulnerable and non-sustainable. It also excludes the farmer from playing the critical role of conserver of genetic diversity, and breeder-innovator in the utilization and development of biodiversity.

The Navdanya program for conservation of the biodiversity of our seeds aims at widening the conservation base to include farmers—the primary custodians of biodiversity.

While efforts at ex situ conservation of genetic resource in high tech, centralized gene banks have been commendable at the national and international levels, conservation at the farmers' level has so far had inadequate attention. Biodiversity conservation at the farmer level differs from *ex situ* conservation measures in two fundamental ways: the former is based on horizontal networking, the latter on a vertical flow. Farmers' conservation involves a two-way flow of biodiversity, while ex situ conservation involves the flow of biodiversity from farmers to gene banks. Conserving biodiversity at the farmer level is necessary for the reasons listed below.

Ecological

Resilience against pests and diseases.

Farm biodiversity acts as an insurance against pests and disease. Cropping patterns based on diverse mixtures of crops reduce vulnerability to disease and pests. Genetic variation in crops also reduces such risks.

Resilience against environmental stress.

Farm biodiversity acts as insurance against drought and climate change. Evolution of diversity in farmers' fields is also necessary as a climate adaptation and climate resilience strategy.

Agricultural change in the Green Revolution paradigm has been based on the intensive use of water and frequent and accurately timed water inputs through irrigation. In periods of droughts and climatic change, these Green Revolution varieties have very high failure rates.

On the other hand, races that have evolved under rain-fed conditions are well adapted to long periods of water stress and variation in the climate. Climate resilient salt and flood tolerant rice varieties have been important for climate resilience and for regeneration of agriculture after climate disasters. In addition, when such varieties are grown in mixtures of as high as nine ("Navdanya") or twelve (*"baranaja"*) crops, the risks of crop failure are further reduced. This insurance is not a trade-off against productivity because when all crop outputs are included in measurements of yield, mixtures generally have higher yields than monocultures.

Biodiversity is essential for providing internal inputs of nutrients and pest control agents on the farm. In sustainable agriculture systems, biodiversity conservation must be an essential part of agricultural production.

The Nutritional Imperative

Biodiversity conservation on the farm is also essential for avoiding nutritional deficiencies in rural communities. Many of the crops threatened with extinction are highly nutritious, though they have been devalued in global food markets. In addition, diverse crops available on the farm make for balanced nutrition. The push for monocultures came from the need to supply cheap food to urban consumers. Consumers are now becoming conscious of both the environmental and health aspects of food. They want food that has not been produced in environmentally degrading ways, and they want food that is nutritious. Biodiversity conservation is therefore a nutritional imperative.

With the triple crisis of malnutrition, farmer's debt, and climate change, new paradigms of ecological breeding are emerging based on participatory and evolutionary breeding.

New research is showing that farmers' varieties have higher nutrition and health benefits than industrial varieties. Industrial breeding and industrial production of wheat based on uniformity, combined with industrial processing which damages the structure of wheat has led to an epidemic of gluten allergies.

Climate change requires constant evolutionary potential in the seed. Today, over six million accessions are conserved worldwide in more than

1,000 gene banks as ex situ germplasm collections, including over 500,000 accessions maintained in field gene banks. Fifteen of the gene banks have long term facilities. About 40% of the accessions conserved in the gene banks are cereals, 15% are pulses, and rest belong to other crops. The lead is provided by the crop based commonwealth Group of International agricultural research center's gene banks. China, India, and the US have the largest gene banks among the countries.

Table 3: Germplasm holding at the ex situ seed Repository of National Gene Bank (as of March 31,1998)

CROP GROUPS	NO. OF ACCESSIONS
Cereals & pseudocereals	70,414
Millets & minor millets	16,587
Oilseeds	24,857
Pulses	26,614
Vegetables & spices	9,284
Released varieties (reference samples)	949
Safety duplicates (IARCs)	9,298
Fiber crops, fiber crops, narcotics	862
M&AP and genetic stocks	415
Others	6,706
Total	1,65,986

National and international efforts at conservation over the last 30 years have concentrated on starting and maintaining ex situ collections, like CGIAR, NBPGR, IBPGR, ICRISAT, and the International Agricultural Research Centres, ignoring the fact that farmers have always conserved varieties successfully using in situ methods. However, exclusive dependence on ex situ cannot conserve biodiversity for the reasons given below. For a truly successful conservation program, both in situ or farm-based conservation has to be given primary importance.

It has been known for centuries now that farmers felt free to deal with issues related to breeding according to their own needs while the formal breeding system has to adapt their strategies to the needs of the commercial system—and this puts strict restrictions on how a plant breeder can work.

Institutional breeding reaches farmers after the seeds have undergone a number of intermediate stages. The stability of the varieties is a main criteria for the formal system, and it is achieved through uniformity, which gives no room for diversity. Institutional breeding is an expensive way of breeding and needs an expensive seed supply infrastructure. For the same reason, it is difficult to breed a number of varieties and, as a result, it becomes essential to breed for wide adaptation. This is done at the cost of compromising the best local adaptation. On the other hand, farmers select seeds for their own use and are therefore concerned about local conditions.

Farmer breeding systems have implications for crop evolution. The two important aspects of crop evolution are "change" and "slowness." Farmers' varieties perform under given growing conditions, climate management and pest regimes, and this will be favored by natural selection. Varieties evolve according to changing land use and modification of cropping patterns. It is so with respect to climate and rainfall patterns as well. A particularly important aspect of this is the co-evolution of crops and their parasites, says Trygve Berg. With our modern systems there is no room for the evolution of varieties. Even when disease resistant varieties are released, the actively evolving populations of parasites sooner or later manage to overcome the resistance. Once this is broken, the diseases spread like wildfire in uniform varieties. The second aspect of evolutionary nature of on-farm breeding is the slowness of progress. This aspect gives farmers time to observe, make adjustments, and absorb the innovations into their farming and food systems. Farmers who select their seeds, use a method called mass selection. This means selection of individual plants according to assessment of performance or appearance. If the farmer selects their seeds from the most fertile patch of the field, it may not be efficient in terms of genetic diversity but will still ensure the highest possible quality with respect to physiological seed development.

2.4 *The Limitations of Gene Banks*

Gene banks are incapable of conserving biodiversity because the philosophy or ideology behind such conservation systems itself is based on three main flaws and inadequacies: scientific flaws, technical inadequacy, and political inadequacy.

A. **Scientific flaws:** The concept of gene banks rests on the assumption that the genetic material of plants (called germplasm by those who accept this theory) can exist independent of the plant itself and that the environment has no role to play in determining or affecting in any manner the characteristics of the variety. This assumption has been proven false. Further, stored in low humidity and at below zero temperatures, the seeds are removed from the process of evolution. Literally frozen in time, the life of the variety will depend upon its ability to adapt to gene bank conditions rather than upon the characteristics which made it worth storing. Thus, the loss of diversity within gene banks is as great as, if not more than, in farmers' fields.

B. **Technical inadequacy:** Ex situ gene banks are totally dependent on high-technology, which is costly, often far beyond the capacity of many developing countries. Technical failure or lack of financial resources can lead to the loss of seed varieties within the gene bank.

C. **Political inadequacy:** Seeds that are stored in gene banks are technically available to farmers, public sector research institutions as well as to the private sector. In practice, individual farmers do not have the same access to these banks—most of which are in developed countries as they lack the required time and resources to get the seed. At the same time, private seed companies with their vast resources and branches in various parts of the world can easily get the seed, and then patent any modification. While the farmer "donates" the germplasm freely, the same is often sold back to him as the property of the company that has patented it.

The same inadequacy exists even on a national level. Establishing and maintaining gene banks is costly. Hence, of the 127 base collections of the International Bureau of Plant Genetic Resources, only 17 are in the national gene banks of developing countries. The latter countries are expected to donate genetic material for conservation purposes to the international gene banks, which are situated mainly in the Global North, and are often supported by privately funded corporations.

The largest gene bank in the world is the US is the National Seed Storage Laboratory (NSSL) in Fort Collins, which stores 232,210 seed samples. Of these, only 64,036 (or 28%) have been found to be healthy. Almost three quarters of the collection have not been tested in at least five years, contain too few seeds to risk testing or do not meet the US standards for viable seeds. Major Goodman, a crop geneticist at the University of North Carolina, has said, "I would maintain that these banks are seed morgues. What goes in, isn't going to come alive."

Why it's Important to Conserve Biodiversity

Given that seed conservation and sustainable agriculture cannot take place in an economic vacuum, but must fit into and transform the economic context in which agriculture is practiced, Navdanya's seed conservation initiatives have, from the very beginning, been associated with building the farmer-to-consumer links so that an economic climate to support sustainable agriculture and seed conservation is created through consumer demand for organic foods.

Navdanya has created more than 137 community seed banks in different parts of India, beginning with seed spread biodiversity based agroecology systems, strengthened farmers' seed sovereignty, food sovereignty, and economic sovereignty throughout the region.

2.5 *Navdanya: A Catalyst for Decentralized Seed Conservation*

Biological diversity cannot be conserved on the basis of centralized, globalized, and hierarchical programs. It needs the building of networks connecting many decentralized initiatives. It has to be based on the logic of "ever widening, never ascending circles" as Gandhi described the philosophy of decentralization. Contrasting this with the hierarchical view of the micro-macro link, his vision was:

"Life will not be a pyramid with the apex sustained by the bottom. But it will be an oceanic circle whose centre will be the individual always ready to perish for the village, the latter ready to perish for the circle of villages till at last the whole becomes one life composed of individuals, never aggressive in their arrogance, but ever humble, sharing the majesty of the oceanic circle of which they are integral units. Therefore, the outermost circumference will not wield power to crush the inner circle but will give strength to all within and will derive its own strength from it."

Reflecting this vision, Navdanya sees its role in seed conservation as a catalyst, creating an ever-widening circle of awareness at many levels from the micro to the macro levels, stepping in to facilitate local groups and communities to take up seed conservation activities, and then stepping out when the local capacities have been built up.

Navdanya's philosophy of community seed banks embodies the seeds of a sustainable living economy from the seed to the table. Farmers save and exchange seeds through the community seed bank network. Seed goes to seed and moves from farmer to farmer. When farmers use their own varieties, they are simultaneously conservers, breeders, and producers. They develop and evolve seeds through breeding and also evolve ecosystems through agroecology. Farmers conserving and shaping fair trade networks enhance their own productivity, incomes, and sustainable livelihoods. Citizens who participate in these networks of biodiverse, organic, and fair trade become co-producers—conserving biodiversity through their conscious decisions on what to eat. In the process, they also improve their health and well-being by eating nutritious, organic, chemical-free food.

2.6 *The Importance of Nutrition*

With agriculture increasingly being viewed as an industry, uniformity of crops through monocultures is becoming an imperative, leading to a loss of diversity. This has generated the paradoxical situation in which plant improvement using diversity as raw material has led to the destruction of the same diversity.

The erosion of this diversity in agriculture was mainly through the manipulation of seeds and plant breeding by scientists in laboratories and not on farms, resulting in the disappearance of traditional crop varieties.

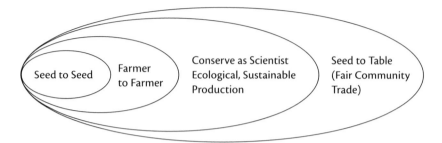

Agriculture shifted to few varieties of wheat and rice derived from a narrow genetic base. The Green Revolution reinforced laboratory-oriented methods of plant breeding to produce high-yielding varieties, hybrid varieties, genetically engineered seeds, and tissue culture.

The central myth that has led to the displacement of diverse farmers' varieties by Green Revolution varieties is that the former are low-yielding, and the latter are high-yielding and have high productivity.

Productivity is basically a ratio between output and input. Farming systems have diverse outputs in terms of diverse crops as well as diverse biomass of the same crop. When the total biomass is taken into account, traditional farming systems based on Indigenous varieties are not found to be low-yielding. In fact, many native varieties have higher yields both in terms of grain output as well as in terms of total biomass output (grain and straw) than the Green Revolution varieties that have been introduced in their place.

The myth of high productivity of the Green Revolution varieties is also not borne out when all inputs are taken into account. Productivity in traditional farming practices has always been high if it is remembered that very few external inputs are required. While the Green Revolution has been projected as having increased productivity in the absolute sense, when resource utilization is taken into account, it has proved to be counterproductive. It has been found that the productivity changes with a shift in emphasis from land to water, from grain to total biomass production (for fodder, straw, fuel, etc.), from weight/acre to nutrition/acre. This is true in respect of industrial inputs like fertilizers.

In terms of efficiency, the Green Revolution technology has proven to be far more inefficient than the technologies it displaced. Whereas in the pre-Green Revolution era, the energy output in terms of food was ten times the input, with the introduction of the Green Revolution, this output has been halved for the same input. Industrial agriculture is equal to the energy input. Again, in terms of finances, productivity has declined with respect to the increasing cost of external inputs like fertilizers and pesticides.

In terms of nutrition, the crops displaced by the Green Revolution technologies include many crops which are better nutritionally than the wheat or rice they gave way to.

Despite its projected success in augmenting wheat and rice output, the Green Revolution is responsible for distortions in the pattern of food production. These distortions have come into being due to the Green Revolution's emphasis on growing a single crop (monocropping). The monocultures of crops and the paddy wheat cycle have wiped out the rich traditional practice of cultivating a variety of crops that once provided both nutritional and economic security to the farmer. The Green Revolution's two major distortions are the increasing disappearance of traditional foods and the loss of nutritional food from our dishes.

Under the pressure of the spread of monocultures of crops, which are traded on worlds markets, highly nutritious crops adapted to local ecosystems and local cultural systems have disappeared. These are often called "lesser-known" or "underexploited crops" because, from the perspective of centralized systems of agricultural development, they are not known and not yet "exploited." However, local communities have known the value and characteristics of these crops for a very long time and have utilized them fully for meeting their nutritional and cultural needs. One of the areas of focus of Navdanya is to prevent the extinction of some of these high-value crops cultivated on a small scale by reintroducing them in farming systems to both increase nutrition and farmers' incomes while conserving resources. We call these the "forgotten foods" of the future due to their nutritional density and climate resilience.

Resource Conserving Nutritious Crops

■ FINGER MILLET (ragi or madua)
Botanical name: *Eleusine corcana*

Finger millet has traditionally been the most important crop grown in many parts of India. The new hybrid crops and the associated agricultural development work have displaced this millet in the last few decades. It is a grain of high nutritive value.

The protein of this millet is as nutritionally rich as milk. It is considered an especially suitable food for diabetic patients. Its malting properties make it special among millets.

> **Nutrients per 100 g of Ragi:** protein 7.3 g, calcium 44 mg, energy 328 cals, phosphorus 283 mg, iron 6.4 mg, carotene 42 mg.

■ FOXTAIL MILLET (kauni)
Botanical name: *Setaria italica*

Foxtail millet is an intermediate drought crop, which gives very high yields. It can grow at elevations up to 6,000 ft. It is often sown as alternate crop with sorghum when rainfall is deficient.

> **Nutrients:** protein 12.3 g, calcium 37 mg, energy 290 cals, phosphorus 280 mg, iron 12.9 mg, carotene 32 mg.

■ BARNYARD MILLET (jhangora)
Botanical name: *Echinochloa frumentaceum*

This is one of the fastest growing millets and can be harvested in approximately four months. The plant has vigorous growth and has wide adaptations in terms of soil and moisture requirements. It is grown for both grain and fodder and is an important forage crop.

> **Nutrients per 100 g of grain:** protein 6.2 g, calcium 20 mg, energy 307 cals, phosphorus 280 mg, iron 2.9 mg, carotene 34 mg.

■ PEARL MILLET (bajra)
Botanical name: *Pennisetum glaucum*

Pearl millet is an important millet in India. It has been cultivated in India and Africa since prehistoric times. The plant grows up to a height of 1.5–1.8 m tall. The grains are gray, rarely yellow in color. Pearl millet is dehusked before consumption. It is usually broken into the rice and cooked or ground into flour.

■ BUCKWHEAT (ogal, phaphra)
Botanical name: *Fagopyrum esculentum*

Buckwheat is a pseudocereal and is usually grown in the Himalayan high altitudes. When tender, the plant is used as a green vegetable. The grain is one of the "phalahar" or foods that can be consumed during fasts. It is available in the plains as "kotu."

■ AMARANTH (marsha, ramdana)
Botanical name: *Amaranthus frumentaceous*

Amaranth is also called "ramdana" or god's grain. Amaranth is the world's most nutritious grain. Its seeds, which come in black, brown, red, gold, and white, can be popped, ground, baked, and cooked. 50–80% of the amaranth plant is edible. Due to its high dry matter content, an equivalent amount of fresh amaranth provides 2–3 times the amount of nutrients found in other vegetables. It has nearly twice as much protein as other cereals, and contains more dietary fiber than wheat, corn, rice, or soybeans, and is a richer source of calcium, iron, and vitamins.

Pulses

Pulse (legume) production has shown no significant gains, with the result that the per capita availability of pulses has sharply declined; and the prices of pulses have escalated to levels beyond the reach of the poor. This has resulted in a sharp decline in protein availability in the diets of those living in poverty. Pulse production has declined rapidly with the spread of wheat and paddy monoculture and is bound to decline further with the policy emphasis on cash cropping and export-oriented agriculture.

Some of the pulses grown organically at Navdanya farm are as follows:

■ BLACKGRAM (urad)
Botanical name: *Phaseolus mungo*

Blackgram is usually grown pure. The large seed variety of blackgram is considered better than the small seeded. It is drought resistant and forms a valuable food resource if millets fail. The proteins in blackgram are comparable more to proteins from animal sources, making this pulse a good substitute for meat.

■ GREENGRAM (moong)
Botanical name: *Phaseolus radiates*

Greengram can be cultivated in both the early and the late monsoon, with varieties specially suitable to each season. Greengram has the least tendency to cause flatulence among the pulses, is easily digestible and is considered to be an ideal food to consume while ill. The flour is often used as a substitute for soap, especially for children.

■ RICE BEAN (navrangi)
Botanical name: *Vigna umbellate*

The resilient rice bean is often grown on marginal and exhausted soil or where other crops do not grow well. An amazing fact about this crop is its ability to be free of pests and diseases. This remarkable bean has an excellent performance from the field to the table. Nutrition wise, as most legumes, it has an excellent profile with 16–25% protein, a good amount of calcium and the presence of many important vitamins, including thiamine, niacin, and riboflavin. It is also a good source of iron and phosphorus. A definite must add to your food basket, it is available on the shelves of Navdanya's organic stores.

▤ COWPEA (lobiya)
Botanical name: *Vigna unguiculata*

Many names exist for *Vigna ungiculata* both in English and vernacular languages. Cowpea was probably first domesticated in the Zambezian region and West Africa with centres of diversity in South and South east Asia.

It exists in a large diversity and has many uses. In India, the dry seeds are mostly consumed as dal or as curries. It is said to tone the spleen-pancreas and stomach meridian; it has a diuretic effect while also relieving conditions such as leucorrhea.

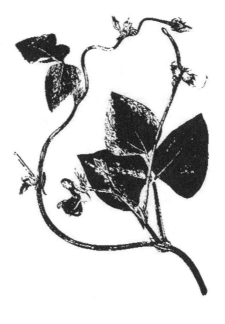

▤ HORSEGRAM (gahath)
Botanical name: *Dolichos biflorus*

A native of the Indian sub-continent, horsegram, has been used as food since 2000 BC. It is prepared in various ways across Garwhal, Maharashtra, and the Kokan area as a dal, a stuffing for rotis, a snack called usal and a curry from the sprouts, with potatoes and onions.

Several health benefits are attributed to horsegram in traditional medicine in India. As per Ayurveda, it has warming properties and should be consumed during the winter months. It may reduce the medha dhatu (body fat), improve sperm count, and regulate the menstrual cycle, to name but a few of its therapeutic uses. One of the main health benefits it is known for is the dissolving of kidney stones.

▓ LENTIL (masoor)
Botanical name: *Lens esculenta*

Lentils or masoor, as we call this legume, is an important component of the diet of several countries, spanning Iran, Ethiopia, several European countries, and India. Here it is eaten as dal, either whole (kala masoor) or split (lal masoor). It is also sometimes made into biryani. One of the easiest legumes to digest, masoor has a high health and taste quotient.

> **One cup of lentils (198 g):** For only 230 calories, one cup of lentils offers 18 g of protein; 1 g of fat; 40 g of carbohydrate (which includes 16 g of fiber and 4 g of sugar); 90% of our daily requirement of folate, 37% iron, 49% manganese, 51% phosphorus, 22% thiamin; 21% potassium, 21% B_6, and 28% B1. Additionally, you get vitamin A, riboflavin, niacin, pantothenic acid, magnesium, zinc, copper, and selenium. It is considered by many as possibly the highest-ranking legume for protein, and since it does not contain sulfur, it produces very little gas.

The Socio-Economic and Political Imperative: Strengthening Farmers' Rights

Farmers have been the conservers and developers of seed since time immemorial. They have a fundamental right to continue to conserve and utilize the biodiversity they have protected.

Farming communities need to redefine words and claim their rights to seed, both as plant breeders and seed suppliers, to regain control over their genetic resources.

The National Consultation on Biodiversity, Sustainable Agriculture, and Farmers' Rights was held in Delhi on March 5 and 6, 1993. Over 70 representatives of various farmers' organizations met and clearly enunciated these rights so that farmers could be free to conserve biodiversity in their fields (see the following box).

🌱 THE RIGHT TO CONSERVE, REPRODUCE, AND MODIFY SEED AND PLANT MATERIAL

Third World farmers are the original donors and custodians of most genetic resources which are the first link in the cycle of food production.

Both patents and sui generis systems create production, distribution and import monopolies. Each time such a monopoly is created, a part of India is handed over to business conglomerates. There are few safeguards against monopolies in either the patents or the sui generis systems. In both cases, the rights of business have been protected over all else.

Article 19 of the Indian Constitution gives every citizen the freedom to practice their respective occupation. This freedom entails the right to shape one's means of production. Since the seed is the primary tool of agricultural production—the "occupation" of farmers—the Article ensures the rights of farmers in their production, reproduction, modification, and selling of seeds. Any changes resulting from IPR regimes would deny farmers their right to freedom of occupation as protected in Article 19 of the Indian Constitution.

We reaffirm our faith in the Indian Patents Act of 1970 which exempts horticulture and agriculture from patentability. We also affirm the farmers right to protect biodiversity on their farms and to reproduce and modify seed freely as non-negotiable.

"Farmers' rights" to biodiversity have been recognized by the FAO. Farmers' rights, as defined in the text of the International Undertaking on Plant Genetic Resources of the FAO, means "rights arising from the past, present and future contributions of farmers in conserving, improving and making available plant genetic resources particularly those in the centres of origin/diversity." However, these rights do not accrue to individual farmers but are rights of states to get aid for biodiversity conservation.

Navdanya started movements to define farmers' rights and responsibilities in India, especially in response to inequitable intellectual property rights regimes which corporations were trying to impose through the TRIPS agreement of the WTO.

We were successful in excluding seeds, plants, and animals from patenting through Article 3(j) of the Patent Act which excludes from patentability "plants and animals in whole or in any part thereof other than microorganisms but including seeds, varieties, and species, and essentially biological processes for production or propagation of plants and animals."

In WTO's TRIPS and UPOV, the rights of breeders are not an obligation. Countries can evolve sui generis laws on plant variety protection. Therefore, breeders' rights on UPOV model do not have to be made national laws.

Dr. Vandana Shiva was appointed to be on the expert group in India that evolved the sui generis law for plant variety protection. India has a law titled Plant Variety Protection and Farmers Rights Act, 2001, which has a clause on Farmers' Rights.

> "[A] farmer shall be deemed to be entitled to save, use, sow, resow, exchange, share or sell his farm produce including seed of a variety protected under this Act in the same manner as he was entitled before the coming into force of this Act."

Farmers' right to seed is connected to three functions: conservation, breeding and production. Corporations define the seed as their invention and intellectual property which prevents farmers from saving and sharing seed. For farmers, seed is a commons to be saved and shared freely among farming communities.

Agenda 21 and the Biodiversity Convention also recognize the need to supplement ex situ conservation with in situ conservation efforts. In situ conservation is critically linked to a strengthening of farmers' rights.

Farmer seed sovereignty is central to addressing farmers' financial and debt crisis. With the entry of corporations in the seed sector, seeds have become a non-renewable commodity for which farmers have to pay a high price. Since the seeds are not bred for local agroecosystems, the rate of crop failure is very high.

The case of failure of Bt cotton hybrids is one example. The Bt cotton needed irrigation and in the semiarid tract of Vidharba, with frequent droughts, it led to frequent failure. The failure of hybrid corn in Bihar is another example.

Seeds are not just the source of life, they are the first means of production for farmers. When seed diversity and sovereignty are eroded, the livelihood and security for farmers is also eroded.

SECTION 3 Soil & Water

3.1 *Understanding and Maintaining Soil Health*

Maintaining soil health is the fundamental principle to successful sustainable organic farming. Poor soil results in unhealthy plants, pests, diseases, and low yields. The key to successful soil health is the correct management of organic matter. This chapter shows how to achieve high yields using a whole systems approach to soil management, including instructions on how to assess a soil's mineral balance.

Soil should have an open friable structure to aid both drainage and water retention which buffers the pH in soils that naturally tend towards acidity or alkalinity. Large, complex organic molecules allow mineral ions to adsorb (stick to) them for later use by the crop. This is very important in areas inundated with periodic heavy rainfall as organic carbon molecules prevent mineral ions from leaching, which keeps nutrients on farm for later use by crops, and prevents the eutrophication of water catchments.

The stable forms of organic matter such as humus and charcoal act as a storage bank and buffer for these minerals. Humic acid, fulvic acid, ulmic acid, and other organic acids from the decay of organic matter, such as carbonic and acetic acids, help make minerals like nitrogen, phosphorous, potassium, and other trace elements bioavailable to plants. These molecules also provide a host for beneficial fungi such as *Trichoderma* and *Penicillium*, which help control pathogens such as *Rhizoctonia*, *Phytophthora*, *Amilleria*, *Pythium*, etc.

It is vital that soil has sufficient plant-available nutrients in the correct balance to ensure that the crop is not deficient and will achieve its maximum genetic potential. The right balance of minerals is essential to the health of soil microorganisms that help as part of the soil food web to produce healthy crops.

This system and its variants have been proven in many countries as a reliable way to achieve high yields. Ensuring soil health is one of the best ways to cultivate plants that are more resistant to pests and diseases, while being robust in adverse conditions such as droughts or heavy rain.

Figure 1: Humus under an electron microscope (*Source*: Rodale Institute)

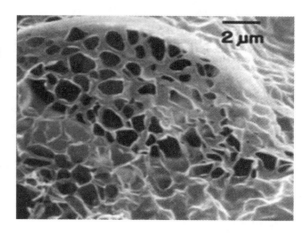

Soil Organic Matter: The Key to Productive Farming

Soil organic matter is one of the most neglected yet most important factors in soil fertility, disease control, water efficiency, and farm productivity.

The combination of dissolved mineral ions in the soil's water is known as a soil solution. The plants absorb these dissolved minerals when they take up this soil solution into their roots to obtain water and nutrients. Yet the importance of soil has become increasingly disregarded as the conventional agriculture industry prefers hydroponic models where plants are directly fed from water solutions. In many agronomy texts, hydroponic methods are seen as the only model for plants to absorb nutrients, and thus the essential role of organic matter in the soil has been deemed irrelevant.

While absorbing minerals through the soil solution is responsible for a significant amount of the minerals that plants need, it is not the only method. Research shows that plants also obtain significantly high levels of nutrients from ion exchange, absorbing larger organic molecules like chelates and amino acids, as well as though direct symbiosis with microorganisms, plant root enzymes, and through the stomata in their leaves. Several of these critical areas of plant nutrition are linked to the organic matter cycles in soils.

The multifunctional benefits of organic matter will be outlined further in this section. The first step is to have an understanding of what constitutes organic matter and, in particular, soil organic matter (SOM).

Soil organic matter (SOM) is very complex. Scientists and researchers are only starting to understand small parts of this complexity. SOM is derived from the excretions and decay of plants, animals, insects, microorganisms, and all biotic life forms. Current research shows that it is composed of two cycles—labile (volatile) and non-labile (stable) fractions—that merge and overlap continuously.

Labile or volatile fraction

The labile fraction is composed of decaying organic matter. This is the most crucial part of the soil organic matter cycles. This is where microbes break down residues of crops, leaves, twigs, branches, root excretions, animal manures, and animal remains and release all the minerals, sugars, and other compounds into the soils to feed plants and other microorganisms. This complex process is known as the soil food web.

The key to this cycle is that it needs to be continuously fed with fresh organic matter to ensure that it is active. In natural ecosystems and under good management, some parts of the decaying organic matter form stable soil carbon and soil organic matter fractions.

Non-labile or stable fraction

The most stable organic matter fractions are humus, glomalin (from fungi), and charcoal (char). Research shows that humus and char can last for thousands of years in the soil. Other fractions are less stable (labile) and can be easily volatilized into CO_2.

Bio-chars (charcoals from living sources) are now being promoted as the stable form of soil carbon along with the benefits that they bring to the soil and to crops.

While bio-chars do have several benefits, the multiple benefits of soil humus are significantly greater, and this manual will concentrate mostly on humus for this reason.

The Soil Food Web

Chemical farming described soil as an empty container, but the soil is a rich ecosystem, rich in biodiversity.

Soil is a living, dynamic ecosystem. Healthy soil is teeming with microscopic and larger organisms that perform many vital functions, including converting dead and decaying matter as well as minerals to plant nutrients. Different soil organisms feed on different organic substrates—their biological activity depends on the organic matter supply.

Nutrient exchanges between organic matter, water, and soil are essential to soil fertility and need to be maintained for sustainable production purposes. Where the soil is exploited for crop production without restoring the organic matter and nutrient contents or maintaining a good structure, the nutrient cycles are broken, soil fertility declines, and the balance in the agroecosystem is destroyed.

Soil organic matter—the product of on-site biological decomposition—affects the chemical and physical properties of the soil and its overall health. Organic matter's composition and breakdown rate affect: the soil structure and porosity, the water infiltration rate and moisture holding capacity of soils, the diversity and biological activity of soil organisms, and plant nutrient availability. Many common agricultural practices, especially plowing, disc-tillage, and vegetation burning, accelerate the decomposition of soil organic matter and leave the soil susceptible to wind and water erosion. However, there are alternative management practices that enhance soil health and allow sustained agricultural productivity. Conservation agriculture encompasses a range of such good practices through combining no-tillage or minimum tillage with a protective crop cover and crop rotations. It maintains surface residues, roots, and soil organic matter, helps control weeds, and enhances soil aggregation and intact large pores, in turn allowing water infiltration and reducing runoff and erosion.

In addition to making plant nutrients available, the diverse soil organisms that thrive in such conditions contribute to pest control and other vital ecological processes. Through combining pasture and fodder species and manure with food and fiber crop production, mixed crop-livestock systems also enhance soil organic matter and soil health. This section recognizes the central role of organic matter in improving soil productivity and outlines promising methods to improve organic matter management for productive and sustainable crop production in the tropics.

Soil organic matter content is a function of organic matter inputs (residues and roots) and litter decomposition. It is related to moisture, temperature and aeration, physical and chemical properties of the soils, as well as bioturbation (mixing by soil macrofauna), leaching by water, and humus stabilization (organo-mineral complexes and aggregates). Land use and management practices also affect soil organic matter.

Farming systems have tended to mine the soil for nutrients and to reduce soil organic matter levels through repetitive harvesting of crops and inadequate efforts to replenish nutrients and restore soil quality. This decline continues until management practices are improved or until a fallow period allows a gradual recovery through natural ecological processes. Only carefully selected diversified cropping systems or well-managed mixed crop-livestock systems are able to maintain a balance in nutrient and organic matter supply and removal.

Soil biodiversity reflects the mix of living organisms in the soil. These organisms interact with one another and with plants and small animals, forming a web of biological activity.

Soil is by far the most biologically diverse part of Earth. The soil food web includes beetles, springtails, mites, worms, spiders, ants, nematodes, fungi, bacteria, and other organisms. These organisms improve the entry and storage of water, resistance to erosion, plant nutrition, and breakdown of organic matter. A wide variety of organisms provides checks and balances to the soil food web through population control, mobility, and survival from season to season.

Building Long-Lasting Soil Organic Matter (SOM)

Humus is the longest-lasting component of soil organic matter. Research has shown that it can last for several thousand years. Over time, biochars will be turned into humus, CO_2, or both, depending on the soil management systems. Humus is generally very resistant to microbial breakdown. However, a combination of synthetic nitrogenous fertilizers and oxidation through poor tillage practices causes it to decline rapidly. Since the top layers of soil usually have the highest percentages of humus, soil erosion is another major cause of humus loss.

Humus is created from a complex mixture of substances from the lignins, oils, and waxes in plants, as opposed to the other main organic compounds, cellulose, sugars, or starches. The exact nature of humus is still being researched.

Under an electron microscope, humus looks like a sponge; it is a sticky substance with numerous porous holes. This is why it can store up to 30 times its own weight in water and why it holds onto soil nutrients and prevents them from being leached out.

The critical issue when building up humus is to allow the ground covers, green manure crops, and stubble to mature to the point where they have formed lignins. The structures of most plants are composed of cellulose and lignin. The lignins are like strong fibers that glue the plant structures together to give flexibility and strength. Young, fresh plants tend to have few lignins as they are primarily made of cellulose, sugars, and starches. The microorganisms readily consume these as food sources, feeding the labile cycle of the soil food web.

Cellulose takes longer to break down. It is formed in plants through building chains of glucose. It is very stable, not water-soluble, and resistant to being degraded. Various microorganisms, especially fungi, can digest it. They use enzymes such as cellulase to break it into glucose and water. Ruminants and termites have symbiotic microorganisms in their digestive tracts that break down cellulose. This is why termites can thrive in arid areas; they get their water and glucose from breaking down the cellulose in wood or grasses.

Soil should be composed of good-quality peds that crumble away into smaller peds when gently squeezed between the thumb and fingers. Organic matter, calcium, clay, microorganisms, air, moisture, and plant roots are needed to build peds. They have interrelated roles, and it is difficult to build good soils without them.

Clays are needed as the binding agents. Nearly all soils have some clay component, including most sandy soils. The regular addition of small amounts of clay will improve sandy soils. However, without organic matter, these clays can be dispersed through the pores in the sand, stopping infiltration and tightly binding to water. Compared to sands, clays have a higher capacity for holding water, making it more difficult for plants to access.

Lignins tend to be the last parts of the plants to be consumed by microorganisms. They can be converted into humus and humic acids, provided that the correct species of microorganisms are available and that destructive farming practices are avoided.

The best way to ensure that plants are rich in lignins is to let them mature and become coarse and woody. It is the lignins that turn tender plants into tough plants. Where possible, let green manures reach full maturity before recycling them into the soil. Young, green manures are great to feed the soil microorganisms with sugars and the crops with nutrients such as nitrogen, but they will not produce as much humus as mature lignified organic matter.

Without regular organic matter inputs, soil organic matter levels can fall over time as the sugars and starches are consumed as the food sources for the soil food web. Soil organic matter tends to volatilize into CO_2 in most conventional farming systems; however, the correct management systems can continuously increase both the non-labile and labile fractions, which means that good organic management systems can increase and maintain the labile fractions as well as the non-labile fractions.

Humus and Soil Nutrients: Ion Exchange

Ions are charged atoms or molecules of minerals. Ion exchange is a significant process in the nutrition of plants in organic systems with high levels of humus. Through ion exchange, plants can separate water into the charged ions they require: the positively charged hydrogen ion and negatively charged hydroxyl ion. The charges on these ions will displace the ions adsorbed on humus to allow them to be absorbed by the plant roots.

Humus can store significantly higher amounts of cations (positively charged ions) and anions (negatively charged ions) than clays due to more sites for ions to adsorb. These sites have positive and negatively charged electrostatic sites that work like magnets to attract the ions.

In standard agronomy texts, the ions are dissolved in the soil solution, and when plants absorb water through their roots, they absorb the dissolved ions as nutrients. Many of these ions also adsorb to the charged sites on humus and will not be dissolved in the soil solution. This prevents the ions from leaching out and causing environmental problems, especially in rivers and seas.

Microorganisms have various essential roles in soil health. Not only are they key in building peds, but their activity also affects the amount of acidity and organic matter in the soil including nutrients and minerals.

Calcium also has a key role. It helps the aggregation of the various components that make up the soil to form into peds. Over time, the microorganisms assemble the various soil particles into peds using humus and calcium. This gives the soil the combination of strength and flexibility. Soil fungi further holds it together with their hyphae (fungal filaments) and secretions like glomalin that reinforce the soil.

Plants are the other important factor. The carbon gifted by the roots feeds the microorganisms that produce the bulk of the organic matter used for humus and building soil. The roots also work as reinforcement and deepen the soil over time. As a result, healthy, living soil will recover from compaction from moderate vehicle use.

🌿 BENEFITS OF HUMUS

- Stores 90–95% of the nitrogen in the soil, 15–80% of phosphorus, and 50–20% of sulfur
- Has many sites that hold minerals, dramatically increasing the amount of plant-available nutrients that the soil can store (TEC)
- Stores cations such as calcium, magnesium, potassium, and all trace elements
- Stores significantly higher amounts of anions (nitrates, sulfur, phosphorous) than clays
- The complex of humic acids (humic, fulvic, ulmic, and others) help make minerals available by dissolving locked-up minerals
- Prevents mineral ions and minerals from being inaccessible
- Prevents nutrient leaching
- Helps to neutralize the pH
- Buffers the soil from drastic changes in pH
- Increases porosity by creating pore spaces for air and water
- Assists with ped formation, which keeps the soil well-drained
- Resists soil erosion
- Encourages macroorganisms (earthworms, beetles, etc.) who increase porosity
- The spaces allow microorganisms to turn the nitrogen in the air into nitrate and ammonia
- Soil carbon dioxide contained in these air spaces increases plant growth
- Helps plant and microbial growth through growth-stimulating compounds
- Helps root growth by making it easy for roots to travel through the soil
- Increases rain absorption through an open structure
- Decreases water loss from runoff
- Can hold up to 30 times their own molecule's weight in water

Soil Horizons (Layers)

Soils generally have three layers or horizons. These are called the topsoil (A) horizon, subsoil (B), and parent material (C).

The most important area is the topsoil, as this is the most fertile zone. It is the area where most of the nutrients and the majority of the crop's feeding roots will be found. The topsoil is formed from the subsoil by the action of crop roots and other parts of plants depositing the organic matter that feeds the soil biology.

The subsoil or B horizon is usually a lighter color as it does not contain the same levels of organic carbon or fertility as the topsoil. Due to poor management, there is either no topsoil or a very little marked difference between these two horizons. These are usually soils of very poor structure and fertility except for some soils of recent volcanic and glacial origin.

The parent material or C horizon is composed of the weathering and decomposing rock material that forms the basis of the soils above. In some cases, this parent material has been deposited by volcanic eruptions as lava or ash, as ground rocks and rock dust from retreating glaciers, or by recent sedimentary events such as flooding rivers, lakes, and sea sands.

Other soils have come from the weathering of harder rocks of igneous, metamorphic, or sedimentary origin over longer geological periods. The standard geological and soil books state that these rocks were weathered down by the normal physical weathering events such as extremes of heat and cold, wind, and running water or by chemical weathering such as weak acids to produce their respective soils. Chemical weathering can be applied to limestone and dolomite. However, most acids and alkalis have little effect on the silicates that form most rocks, especially the very dilute organic acids that are formed in normal soil processes.

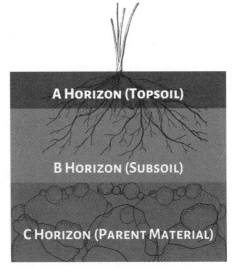

Figure 2: Soil layers

A Horizon (Topsoil)

B Horizon (Subsoil)

C Horizon (Parent Material)

Similarly, the extremes of atmospheric weather such as abrasive winds, very hot or freezing cold temperatures and fast-flowing water rarely reach the C horizon of the soil and do have enough effect to cause major weathering at this depth.

Forms of Weathering

Scientists have found that a wide range of organisms and plant roots mine soil for specific minerals, which leads to three very important processes.

1. Orphaning of other minerals

When the biological agents extract the mineral that they want (ex. potassium), they will "orphan," release the other mineral(s) it was attached to (ex. silica). Other organisms can take these orphaned minerals or combine with other minerals to form new compounds outside of the parent rock.

2. Decomposition of rocks

The gradual loss of the minerals that are holding the parents rocks together causes them to crumble into smaller particles and start the process of forming the physical basis of the soil type.

3. Feeding the soil food web

The initial microorganisms that mine the minerals continuously die and are consumed by other organisms to form a food chain through the soil food web. This results in some of the newly mined minerals being taken up by plant roots.

Other forms of biological weathering

Biological activity is the major factor in the decomposition of rocks that form soil. Plants roots and worms are significant causes of biological weathering. There are many studies showing the ability of plant roots to extract significant amounts of minerals and weather rocks. Recently, studies have shown that deep rooted plants can extract nitrogen and other nutrients from the parent rocks.

Smaller rock particles are weathered when they go through the digestive tracks of worms and these minerals are bio-accumulated in the worm casts in the top soil, or A horizon.

The ability of microbes to extract significant amounts of minerals from the parent rock has been successfully used for commercially viable mining operations to collect minerals such as copper and gold. Mining companies have isolated specific microbes that they can use to inoculate the rocks to extract the required minerals in commercially viable quantities.

Soil Biodiversity

Worms are used to improve soil quality. Soil biodiversity reflects the variability among living organisms, including a myriad of organisms not visible with the naked eye, such as microorganisms (e.g., bacteria, fungi, protozoa, and nematodes) and meso-fauna (e.g., acari and springtails), as well as the more familiar macro-fauna (e.g., earthworms and termites). Plant roots can also be considered as soil organisms in view of their symbiotic relationships and interactions with other soil components.

These diverse organisms interact with one another and with various plants and animals in the ecosystem, forming a complex web of biological activity. Soil organisms contribute a wide range of essential services to the sustainable function of all ecosystems. They act as the primary driving agents of nutrient cycling, regulate the dynamics of soil organic matter, aid in soil carbon sequestration, modify soil physical structure, increase water infiltration, and enhance the efficiency of nutrient acquisition.

These services are not only essential to the functioning of natural ecosystems but constitute an important resource for the sustainable management of agricultural systems.

Figures on Soil Biodiversity

Nowhere in nature are species so densely packed as in soil communities. Soil biodiversity is characterized by:

+ Over 1,000 species of invertebrates may be found in a single m² of forest soils.
+ Many of the world's terrestrial insect species are soil dwellers for at least some stage of their life cycle.
+ A single gram of soil may contain millions of individuals and several thousand species of bacteria.
+ A typical, healthy soil might contain several species of vertebrate animals, several species of earthworms, 20–30 species of mites, 50–100 species of insects, tens of species of nematodes, hundreds of species of fungi, and perhaps thousands of species of bacteria and actinomycetes.

Relative number of microbes in a handful of soil (100–200 grams)

Bacteria	50 billion
Actinomycetes	2 billion
Fungus	100 million
Protozoa	50 million
Nematodes	10 thousand
Arthropods	1 thousand
Earthworms	0–2

Soil is composed of air (25%), water (25%), minerals (45%), and organic matter (5%) which interact with each other to sustain plants on planet Earth. This is possible through the interactions of millions of living macroorganisms and microorganisms (microbes, fungi, bacteria, earthworms, ants, etc.). These organisms interact with each other by decomposing dead matter and breaking down complex nonliving mineral parts of the soil to form a natural system.

Soil on the Navdanya farm demonstrates a unique natural balance in the living and nonliving entities. The soil samples collected from various Navdanya crop fields showed higher organic matter in comparison to chemical farms. Microbes such as bacteria and fungi are higher in the soil on the Navdanya farm than the chemically-managed soils from neighboring farms. A plethora of organisms can be found in the soil at the Navdanya farm.

A detailed study on the soil was carried out in a group of crops grown on the Navdanya organic farm and surrounding chemical farms showed that soil from organic farms were richer in the biological and physico-chemical parameters than chemical farms. The control site was barren land where no crops were grown. The process of the soil biology mining the parent material for new minerals is the key to maintaining a fertile and productive soil. This can be achieved if the soil biology is active and healthy. The key to healthy soil biology is to feed it with organic matter.

The other critical element is the presence of deep-rooted plants. The cropping system needs to include species of plants that can send their roots down to the C horizon and extract the minerals. These plants also have the other critical role of opening up the subsoils to allow the infiltration of air and water as well as the deposition of organic carbon that is shed by roots.

Ideally, a farming system should not be exporting more nutrients from the soil, through the sending crops off-farm, than can be replenished from the parent rock through the farming system. It is critical to look at two key factors:

1. The mineral and nutrients being depleted.
2. The minerals and nutrients being replenished.

Improving the soil can be done at the same rate or through replenishment after the crop is harvested. If neither of these are being done, then the land management system is unsustainably degrading the soil.

Table 1: Fungi (CFU × 10^3/g) population under different crops and farming practices

CROPS	CONTROL*	NO INPUT	CHEMICAL FARMING	ORGANIC FARMING
Wheat	5.5	22.0	20.0	66.5
Potato	3.5	7.0	6.0	120.0
Garlic	7.0	20.0	19.5	94.0
Mustard	3.0	7.2	11.5	111.0
Chickpea	6.5	18.7	8.0	180.0
Chili	8.5	20.0	14.0	160.0
Pumpkin	7.0	14.1	12.0	52.0
LSD (=0.5)	4.2	7.3	6.5	11.1

Table 2: Fungi (CFU × 105/g) population under different crops and farming practices

CROPS	CONTROL*	NO INPUT	CHEMICAL FARMING	ORGANIC FARMING
Wheat	2.5	4.4	4.0	15.0
Potato	3.0	8.4	8.0	12.0
Garlic	4.5	10.4	7.0	26.0
Mustard	3.5	6.0	4.0	10.0
Chickpea	5.0	9.3	7.0	14.0
Chili	2.0	5.8	5.5	12.5
Pumpkin	4.0	8.8	8.0	29.0
LSD (=0.5)	1.7	1.9	1.8	3.5

Table 3: Organic matter (%) content under different crops and farming practices

CROPS	CONTROL*	NO INPUT	CHEMICAL FARMING	ORGANIC FARMING
Wheat	0.8	1.2	1.14	1.64
Potato	0.8	0.86	0.74	1.27
Garlic	0.85	1.19	1.17	2.21
Mustard	1.12	1.35	1.34	2.68
Chickpea	0.9	1.17	1.12	1.47
Chili	0.92	0.97	0.95	1.62
Pumpkin	0.85	0.93	0.85	1.29
LSD (=0.5)	0.11	0.18	0.15	0.21

Table 4: Total Nitrogen (N) (%) under different crops and farming practices

CROPS	CONTROL*	NO INPUT	CHEMICAL FARMING	ORGANIC FARMING
Wheat	0.08	0.11	0.11	0.16
Potato	0.08	0.09	0.07	0.13
Garlic	0.09	0.11	0.11	0.22
Mustard	0.11	0.14	0.13	0.26

Chickpea	0.09	0.11	0.11	0.14
Chili	0.09	0.1	0.1	0.16
Pumpkin	0.09	0.09	0.09	0.13
LSD (=0.5)	0.03	0.02	0.04	0.05

Table 5: Available Potassium (K) (mg/kg) under different crops and farming practices

CROPS	CONTROL*	NO INPUT	CHEMICAL FARMING	ORGANIC FARMING
Wheat	124.3	115.6	106.6	137.3
Potato	110.9	120.3	141.7	141.4
Garlic	108.8	95.0	73.9	175.2
Mustard	112.0	118.7	117.7	135.5
Chickpea	110.0	115.0	114.0	132.2
Chili	108.0	102.6	100.5	123.6
Pumpkin	105.0	102.8	142.3	140.5

Standard Agronomy has downplayed the role of organic matter in plant nutrition and consequently ignored its critical and multifunctional role in helping plants to obtain sufficient levels of minerals, reducing pests, and creating resistances to diseases.

The term and the concept of the rhizosphere were proposed by the German scientist Lorenz Hiltner in 1904. Hiltner observed that the greatest concentration of soil microorganisms could be found in a narrow zone surrounding the roots of plants. He also observed that they were feeding on the sheaths that roots shed as they grow, as well as a number of other exudates such as sugars and amino acids.

He proposed that the overall health of plants depended on the health of these diverse colonies of microbes in that they helped to prevent diseases and assisted with the uptake of minerals. Based on his observations, he hypothesized that "the resistance of plants towards pathogenesis is dependent on the composition of the rhizosphere microflora." He claimed that the quality of the plants grown may be dependent on the composition of the root microflora.

Numerous studies show that these microbes produce a range of compounds that plants use for nutrition. The best known are the *Rhizobium* bacteria that live in the roots of legumes. These organisms convert nitrogen into forms that plants can use.

Other examples of groups of beneficial microorganisms are the vesicular-arbuscular mycorrhiza (VAM) and related fungi. These fungi live in the roots of plants and extend their threads of mycelium into the soil to mine minerals. They exchange these minerals for glucose. They are particularly important with the uptake of phosphorous in many plant species as they have enzymes that can split phosphorous off rocks and reach locked-up molecules such as iron phosphides and tri-calcium phosphates, which they feed into plant roots.

Many of these fungi also protect their hosts from diseases as well as helping them with nutrition. The science around the rhizosphere has now increased significantly. However, the complexity of the interactions of the massive biodiversity in the soils around the roots means that it is still not well understood and as a result, it is not being widely applied in most agriculture.

One of the critical issues emerging is that high soil microbe biodiversity is essential so that these microorganisms can work in symbiosis to fight off pathogens. The most recent study into disease suppressive soils found that more than 33,000 species worked together to suppress diseases. Similarly, the number and types of free-living microorganism species that fix nitrogen continues to increase as researchers find more.

Most texts will only mention *Rhizobium* bacteria that live in symbiosis in the nodules of legumes, though there are also the free-living nitrogen-fixing organisms such as *Azotobacter*, cyanobacteria, *Nitrosomas*, and *Nitrobacter*. Even more live in the rhizosphere and help plants take up nitrogen from the soil. Researchers are just starting to discover them.

Chemical and Synthetic Fertilizers

Synthetic chemical fertilizers are the main method of providing nutrients to conventional farming. In many cases, they are the only nutrient input in the farming system. At first, they were found to give significant increases in yield; however, this advantage is disappearing over time, and yields are starting to decline.

How Microorganisms Assist Plants

Soil organic matter is the key to a healthy soil biology. The amount of biological activity in a soil is directly related to levels of soil organic matter.

Make nutrients available

+ Decompose organic matter and release nutrients
+ Dissolve minerals from rock
+ Produce chelating and complexing nutrients
+ Improve soil structure
+ Build peds by disturbing and stirring clay and other particles into open random forms and gluing them together with humus, organic polymers, and fungi hyphae
+ Moves soil particles around and increases porosity

Interacts with plants

+ Predating pathogens (e.g., eating pests and diseases)
+ Protozoa eating bacteria wilt
+ Fungi eating nematodes
+ Produce antibiotics that kill pathogens
+ Suppressing pathogens through outnumbering them
+ Detoxifying synthetic chemicals and poisons
+ Fixing nutrients like nitrogen from the soil air into plant-available forms (such as Azobacter and cyanobacteria)
+ Create enzymes, vitamins, and amino acids
+ Fixes soil nitrogen into plant usable forms (rhizobia)
+ Directly feeds nutrients into plants (VAM)

The United Nations MA Report and other research on the environment found that loss of soil fertility is resulting in yield decline around the world. Farmers need to dramatically increase the amounts of synthetic fertilizers and pesticides to maintain yields, and this is causing major environmental problems.

> "Since 1960, flows of reactive nitrogen in terrestrial ecosystems have doubled, and flows of phosphorus have tripled. More than half of all the synthetic nitrogen fertilizer...ever used on the planet has been used since 1985." (UN 2005)

Soluble fertilizers from conventional farming systems are causing the eutrophication of freshwater and coastal marine ecosystems and acidification of freshwater and terrestrial ecosystems. These are regularly creating harmful algal blooms and leading to the formation of oxygen-depleted zones that kill animal and plant life. The dead zones in the Gulf of Mexico and parts of the Mediterranean are caused by this. The increased frequencies of toxic blue algae bloom in rivers and estuaries are attributed to the increased nitrogen and phosphorous run-off from current land management practices. Scientists have shown that tropical and subtropical oceans are acutely vulnerable to nitrogen pollution. They stated: "Our findings highlight the present and future vulnerability of these ecosystems to agricultural run-off."

In Australia, this is already occurring in the Great Barrier Reef, the world's largest and most biodiverse tropical reef system, due to the run-off of nutrients from farms causing damage. This increase in nitrogen and phosphorous is causing algal growths that cover the corals, preventing them from accessing nutrients and sunlight for photosynthesis, causing deaths and declines in their population. This is particularly true with the coastal fringing reefs. Nearly all of the coral fringing reefs that are adjacent to farming regions in Queensland died in the 20th century. Farming in Europe is resulting in significant nitrogen contamination of water catchments. According to the researchers:

> "Nitrogen contamination of ground and surface water in the Seine, Somme and Scheldt watersheds, as well as in the receiving coastal marine zones, results in severe ecological problems."

The researchers looked at a range of strategies in catchments off the Southern Bight of the North Sea to reduce nitrogen run-off based on several scenarios to improve Good Agricultural Practice (GAP).

> "Previous results showed that the implementation of classical management measures involving improvement of wastewater purification and 'good agricultural practices' are not sufficient to obviate these problems."

GAP systems originated in Europe and are designed to ensure that conventional agriculture does not damage the environment or cause health problems to people who consume its food. They are held up as examples of best practice even though they are labeled as "good" practice. Researchers in Europe have found that despite adopting better practices than those prescribed under GAP systems, the reductions in pollution were too small to make a significant change to the damage caused by nitrogen fertilizers. Adding organic matter to the soil is the most effective way to increase soil quality. This can be done in various ways, including using mulches, both living and dead, and green manure cover crops.

"However, only an overall 14–23% reduction in N could be achieved at the outlet of the three basins, by combining improved wastewater treatment and land use with management measures aimed at regulating agricultural practices. Nonetheless, in spite of these efforts, N will still be exported in large excess with respect to the equilibrium defined by the Redfield ratios, even in the most optimistic hypothesis describing the long-term response of groundwater nitrate concentrations." (Thieu et al. 2009)

In a follow up study, the researchers found that adopting organic management systems would significantly reduce the problems of aquatic eutrophication.

"It leads to a significant reduction of agricultural production that finally brings the three basins closer to autotrophy/heterotrophy equilibrium. Nitrate concentrations in most of the drainage network would drop below the threshold of 2.25 mg N/l in the most optimistic hypothesis. The excess of nitrogen over silica (with respect to the requirements of marine diatoms) delivered into the coastal zones would be decreased by a factor from 2 to 5, thus strongly reducing, but not entirely eliminating the potential for marine eutrophication." (Thieu et al. 2010)

These studies confirm the results of earlier research studies from North America and Europe showing that organic systems are more efficient in using nitrogen than conventional farming systems. Significantly, because of this efficiency, very little nitrogen leaves the farms as greenhouse gases or as nitrate that pollutes aquatic systems.

The governments of Germany and France have encouraged conversion to organic farming to improve water quality, particularly in relation to its nitrogen and pesticide content.

Synthetic chemical fertilizers are significant contributors to climate change in terms of the energy used to manufacture them and their contribution to nitrous oxide (N_2O) and methane (NH_4).

One of the most significant greenhouse gases emitted by agriculture is nitrous oxide (N_2O). One N_2O molecule is equivalent to 310 CO_2 molecules in its greenhouse effect in the atmosphere. It has a mean residence time in the atmosphere of 120–150 years and also contributes to the depletion of the ozone layer in the atmosphere.

The biggest contributor to human-produced (anthropogenic) N_2O pollution is the use of synthetic nitrogen fertilizers such as urea and ammonium nitrate in conventional agriculture. It is even higher when all the CO_2 and N_2O that is emitted in the production of these energy-intensive fertilizers are included in the totals.

Most governments do not factor the CO_2 emissions that result from the production of these synthetic fertilizers into the greenhouse gas levels caused by agriculture. These emissions are usually factored into the emissions of the manufacturing sector, even though the primary reason for manufacturing them is agriculture. N_2O is expected to become an even greater issue since the increasing levels of this gas causes more damage to the ozone layer than the more commonly known chlorofluorocarbons (CFCs). The researchers showed that nitrous oxide is the single most important ozone-depleting substance (Ravishankara 2009).

Soil Management

The most critical issue in developing fertilizing systems in organic agriculture is to have a farming system that is continuously incorporating organic matter into the soil. This can be achieved through numerous methods, such as using green manure cover crops, incorporating crop residues, and adding compost. It is important that these systems actively stimulate the two organic matter cycles: labile and non-labile. The labile cycle is critical to the fresh release of nutrients to the crop; the non-labile cycle is critical in building the stable soil organic matter that holds most of the nitrogen and other nutrients needed by plants.

Using Microorganisms to Convert Organic Matter into Stable Forms

The stable forms of soil organic matter such as humus and glomalin are manufactured by microorganisms. They convert the carbon compounds read-ily oxidized into CO_2 into stable polymers that can last thousands of years in the soil.

Some of the current conventional farming techniques, such as adding synthetic nitrogen fertilizers and incorrect tillage, result in the soil carbon deposited by plant roots being oxidized and converted back into carbon diox-ide. This is the reason why soil organic matter (carbon) levels continue to decline in these farming systems.

The other significant depositories of carbon are soil organisms. Research shows that they form a considerable percentage of soil carbon. It is essential to manage the soil to maintain high levels of soil organisms.

It is essential that farming techniques stimulate the species of soil micro-organisms that create stable carbons, rather than only stimulating the species that consume carbon and convert it into CO_2.

Creating Stable Soil Organic Matter (SOM)

The process of making composts uses microbes to build humus and other stable carbons. The microorganisms that create compost continue working in the soil after compost applications, converting the carbon gifted by plant roots into stable forms. Regular applications of compost or compost teas will inoculate the soil with beneficial organisms that build humus and other long-lasting carbon polymers. Over time, these species will predominate over the species that chew up carbon into CO_2. Regular applications of composts or compost tea also increase the number and diversity of species living in the soil biomass. This ensures that a significant proportion of soil carbon is stored in living species that will make minerals available and protect the health of the plants.

Compost and Compost Teas

Research shows that good quality compost is one of the most important ways to improve soil. It is very important to understand that compost is a lot more than a fertilizer. Compost contains humus, humic acids, and most importantly, a large number of beneficial microorganisms that have a major role in the process of building healthy soils, especially humus.

Compost is the ideal way to improve soil quality, build soil organic matter levels, and correct mineral imbalances. The best way to balance the minerals in soil is to work out the amounts needed through a soil test and add the depleted ground minerals, such as rock phosphate, ground basalt, potassium sulphate, gypsum, etc., into the compost material when starting a compost pile. The biological processes that form compost will make these minerals readily available to plants in both quick-release and slow-release forms. The resulting mineral rich compost should be spread around the crops and periodically, trace elements can be applied. The trace elements can be mixed with molasses, compost tea, or both, and brewed for several days to make them bioavailable.

Compost teas have been successfully used to inoculate soils with beneficial microorganisms that will increase soil carbon, improve soil quality, and in many cases, suppress soil and plant diseases. Compost teas are made by adding small amounts of compost to water and brewing for a while to ensure that the microorganisms are active. These can be sprayed out evenly over the field. This system ensures that the soil's biological activity releases a steady flow of all the nutrients needed by the crop to produce a good yield. The complete nature of the nutrition program ensures that there are no deficiencies. It is best to spray out the teas in the late afternoon so that the microorganisms are not killed by ultraviolet light.

Composting Methods

There are many methods that can be used to make compost. The following instructions will provide an overview.

Sheet composting

Fresh manure is spread over a cover crop or crop residue and the composting process occurs in soil. It is usually a requirement of this system that a green manure crop is grown afterwards, which is either slashed or plowed into the soil. One advantage is that very little nutrients are lost through leaching or volatilization.

The risk is residual chemicals in manure such as drenches, pesticides, atrazine, or antibiotics can interfere with the microbial breakdown of the raw organic matter and of weed seeds germinating.

Aerobic composting

The advantage of this method is that it is the fastest way to make compost. The disadvantages are that additional labor is required for the regular turnings, and each turning results in the loss of volatile nitrogen and other compounds.

+ Ideal carbon to nitrogen (C:N) ratio of 25–35:1.
+ Moisture 60% at point of making (when squeezed hard, moisture appears on outside of the bolus).
+ Temperatures that reach up to 70°F.
+ Constant supply of oxygen by turning at least weekly.
+ Well-mixed.
+ Piles up to 2 meters high with a 45–60 degree slump angle.
+ Management of high pH rock dusts such as lime and dolomite to monitor nitrogen losses.

Anaerobic compost

+ As above for aerobic compost.
+ Less oxygen means that it takes more than twice as long before it is ready to use.
+ Less nitrogen loss.
+ Anaerobic bacteria create a range of low pH organic acids and enzymes that are useful in making mineral rock dusts (lime, rock phosphate, crushed basalt, dolomite, gypsum, etc.) bioavailable.
+ Cheaper to make due to less costs for turning to oxygenate.

The Permanent Compost Pile

One of the easiest ways to make compost is to have a permanent pile that is continuously fed with all forms of organic matter. The pile is started with a combination of fresh and dried organic matter with the addition of local worms. All the sources of organic matter from the farm and around the farm are continuously added to the pile so that it is fed at least every week. The sources can include old palm fronds, branches, leaves, food scraps, weeds, animal carcasses, manure, and any other form of organic matter.

Over time, this heap will have multiple species of worms, fungi, bacteria, and other beneficial microorganisms that will break down the organic matter into humus-rich compost.

The heap is periodically opened up and the humus-rich compost can be collected for use on the fields. The semi-decayed and undecayed organic matter is separated from the compost and left on the same site to ensure that the heap is still continuously making compost. Since this heap continues to be fed, many people use this method and can sustain compost heaps that are over 20 years old, which continuously produce good quality compost.

Vermi-Compost

Compost can be made from any organic matter sources. This includes animal manures, grass, bushes, branches, leaves, some weeds, and overgrown vegetation.

Most farmers become good harvesters of organic materials from diverse sources. Letting the vegetation regenerate around the farm on hillsides, gullies, creeks, and along the field borders is the best ways to ensure a constant supply of organic matter for compost making. This can be regularly managed to prevent it from getting out of control and the harvested cuttings can be used for making compost.

Brown and Green Sources

Many compost books will talk about having the materials in an ideal C:N ratio.

They will also have a table of the carbon to nitrogen ratios of many ingredients and examples of how to do the mathematics that are needed to work out the percentages of each of these to make the ideal ratio when using multiple sources.

Most farmers find this too complicated to use. A more practical way is to think of using a mixture of brown and green organic matter sources. Brown or dried organic matter sources are usually high in carbon and low in nitrogen. Green or fresh organic matter sources, such as freshly cut grasses, usually have high levels of nitrogen in relation to carbon.

Mixing the brown and the green will give a good ratio of carbon to nitrogen. Experience will be the best guide to getting a good result.

Time is the important factor. Lower levels of nitrogen will mean that it will take longer to break down into humus rich compost. This means that farmers should start making compost at least six months to a year before they need to use it.

Lowering Greenhouse Gases with Compost

In some parts of Europe and North America, up to 10% subsoil clay is added to improve the texture. An acidic clay will stop the volatilization of nitrogen as ammonia. Ammonium ions will stick to the clay. This lowers the amount of nitrogen-based greenhouse gases escaping from the compost.

❦ THE BENEFITS OF COMPOST

° Adds humus and organic matter to the soil.

° Inoculates soil with humus-building microorganisms.

° Improves soil structure to allow better infiltration of air and water.

° Humus stores between 20–30 times its weight in water and significantly increases the soil's capacity to store water.

° Humus stores nitrogen and other nutrients for later use by plants.

° Provides a variety of instant and slow-release nutrients.

° Supplies a large range of beneficial fungi, bacteria, and other useful species.

° Suppresses soil pathogens.

° Fixes nitrogen.

° Increases soil carbon.

° Releases locked-up soil minerals.

° Detoxifies poisons.

° Feeds plants and soil life.

° Builds soil structure.

❦ CASE STUDY ❦

In Tigray, yields were more than double, there was better disease resistance, and water use efficiency when compost was added. During an outbreak of Stripe Rust, the composted wheat did not get effected by the disease and gave yields of 6.5 tons per hectare whereas the uncomposted fields had to be sprayed with fungicides and still only yielded 1.6 tons per hectare. Additionally, the composted fields used 30% less water and were more drought tolerant.

Similarly, it has been shown that potassium tends to leach out of compost heaps. Clay platelets tend to strongly attract potassium ions and will prevent leaching.

Anaerobic composts can have a carbon-nitrogen ratio of more than 30:1 as the longer time favors more fungi than bacteria. Fungi need less nitrogen to do the job of breaking down the raw materials. A higher carbon ratio means that it will take longer for the microorganisms to break down the organic matter and turn it into humus; however, it will lessen the nitrogen loss and result in a compost with more useable nitrogen for the crop. This also lowers the amount of nitrogen-based greenhouse gases escaping from the heap.

Covering the compost pile with black plastic or fresh subsoil clay helps keep the moisture level stable. The black plastic will solarize the weed seeds on the surface of the pile while the compost heat will destroy most of the internal weed seeds. Please note that some seeds, especially those with hard coats, can survive and germinate later.

Recent research from the US has found that covering composts with a deep layer of wood chips stops all emissions of volatile organic compounds (VOCs), including greenhouse gases.

Worms can be added to composts when the compost heap starts cooling down. They are particularly beneficial to anaerobic composts because, over time, they will turn over and aerate the whole heap. For farm-scale compost piles, this saves hours of work and many liters of diesel.

Every time a compost heap is manually turned, it releases a range of greenhouse gases. Covering it with plastic, clay, or wood chips and letting worms slowly aerate the pile significantly lowers the amount of greenhouse gases such as CO_2, methane, ammonium, and nitrous oxide (N_2O). These are reabsorbed and biologically degraded by the microorganisms, turning them into more stable organic carbon and nitrogen forms, such as humus and amino acids.

Biodynamic Preparations

Biodynamic preparations such as horn manure (also called "500") can work in a similar manner to compost teas and have been very successful in building soil organic matter, especially humus.

Green Manure: Cover Crops

Green manures are crops that are grown purely to improve soil health and fertility by adding in fresh organic matter and nutrients, such as nitrogen, when they are incorporated in the soil.

Green manures are generally a part of a crop rotation that is used to break the weed and disease cycles. These multifunctional benefits are explained in greater detail in "Section 5: Biodiversity for Pest Control: Managing Pests without Pesticides." The other main reason for green manure is to plant and then incorporate them into the soil just before the cash crops to provide a release of nutrients for the cash crop.

Virtually all plants can be used as green manures; however, legumes are the preferred plants as they can provide significant amounts of both organic matter and nitrogen. The use of green manures is one of the oldest and most proven methods to improve nitrogen and organic matter levels in the soil.

Traditional Methods of Composting

Farming activities generate a lot of wet and dry biomass. This biomass is indiscriminately burnt, and the energy and nutrients are lost to the atmosphere, which pollutes the air. However, the energy and nutrients in the biomass can be sustainably used and restored to the soil that is the actual source. Biomass can be converted to compost that is easily taken up by the soil and used by the living system in it. They break down the compost particles into finer nutrients that are available to the plants. There are various methods of preparing different composts.

Farmyard manure (FYM)

Cow or animal manure can be used as farmyard manure. The manure must be stored under shade for three months to retain moisture in it. The farmyard manures provide nutrients in the right proportion.

▓ Pit Method

The crop residues and biomass available on the farm can be composted by using this method.

Ingredients:

Dry twigs	Soil
Cow manure	Green biomass
Cow urine	Dry biomass

Method: Dig a pit measuring 10 ft long, 3 ft wide, and 3 ft deep, preferably close to the field. Place the dry twigs at the bottom of the pit so that they provide spaces for aeration. Follow this by spreading the cow manure on top; it should occupy 25% of space. Above this, fill the green biomass followed by dried biomass. The heap should be raised ½ ft above the ground.

After completing this process, the pit should be sealed with cow manure in a diluted form using cow urine and soil. The proportion of cow manure, green biomass, and dried leaves is 25:25:50.

After a week's time, the heap of biomass becomes contracted due to decomposition of organic matter from microbial activity. Add more biomass and seal the heap by using cow manure, cow urine, and soil.

Final Product: The compost will be ready to use in approximately three months. The compost will have a brown color with a pleasant aroma.

■ GHANAMRUTH

In this method, cow manure is transformed into compost. The rationale behind this is to multiply the number of microorganisms to hasten the process of decomposition.

Ingredients (for one acre):
200 kg cow manure
2 kg lentil flour
2 kg jaggery
½ kg fertile soil
10 L cow urine

Method: Spread the cow manure on the ground, and layer the lentil flour on top of it. Add the jaggery to the mixture. Now sprinkle the cow urine over it and add soil to this mixture. Mix everything thoroughly by stamping on it. After it is thoroughly mixed, spread the mixture further. The thickness of the cake must be 2–3 inches thick.

Final Product: The cake must be allowed to dry for 20–25 days and should be covered with gunny bags if the temperature is high. After seven days of preparation, cavities can be seen on the cow manure cake, which is an indicator of microbial activity in the mixture. Once it is completely dried, the cake can be cut into small pieces and stored in gunny bags for six months.

Care must be taken to keep the cakes moistened with water periodically.

Application: At the time of irrigation to the field, the bag containing cakes must be placed at the source of the irrigation. This will facilitate application of compost throughout the field. It can also be applied at the time of sowing or weeding.

Liquid Nutrients through Fermentation

■ Jeevamrit

Ingredients required for one acre:

200 L water

10 kg cow manure

10 L cow urine

2 kg legume flour

2 kg jagger or molasses

1 fistful fertile soil (as inoculant)

Method of Preparation: Use a 250 L plastic drum with a tap and lid. Add the ingredients in the same order as listed above. Close the drum with a cloth and place the lid tightly. To allow oxygen for microbial activity into the mixture, stir the contents of the drum clockwise and counter-clockwise three times a day. This encourages rapid multiplication of microbes. At the end of the third day (in hot and humid conditions) the liquid will be ready for use in the field.

Rationale: This mixture helps to increase the soil microorganism abundance and diversity. Microbes get their nutrients from the provided mixture. This is not a fungal preparation of fertilizer. The mixture is most effective on the third day, beyond which the microbial population declines as they die and degenerate. It can't be stored beyond four days during summer and/or in humid conditions. When microbes die, they act as fertilizer and contribute carbon, potash, and sulfur.

Application: The liquid must be applied on the third day of its preparation. The field should have sufficient moisture. When irrigating the field, open the tap of the drum and allow the *Jeevamrit* mixture to flow with water, which ensures uniform distribution in the entire field. Upon reaching the soil, the microorganisms will multiply and utilize the available carbon.

Stages of application:

1. At the time of plowing to make the soil friable.
2. At the time of tillering (25–30 days after germination).
3. Before flowering, seeding, and fruiting.

Limitation: If the crop is deficient of nitrogen, *Jeevamrit* will not work. It works in the field where organic matter is present profusely.

■ GARBAGE ENZYME (NUTRIENT)

Ingredients (10 L preparation)

3 kg kitchen and fruit wastes.

1 kg molasses or jaggery (if substituting grape use 2 kg)

10 L water

The mixture should consist of equal amounts of rotten fruits and vegetables. Avoid potato and starchy food as they emit a foul smell. Ripened papaya and sour fruits such as orange, lemon, and tomato are a good combination. Cut them into small pieces. Rose flowers can be added to remove the foul smell.

Method of Preparation: Add the ingredients to a plastic drum (preferred). Close the drum with a lid and write on it the date of preparation. Every 15 days, open the lid and stir it. After three months, the mixture is ready to use. Write the date on which it was ready to use.

Rationale: It is a fermentation process and destroys the microbes. The solution acquires a pH-4. It is rich in acetic acid. The garbage enzyme can be kept for three years.

Application: Garbage enzyme should be used in a proportion of 1 ml to 1 L of water. Spray during the morning and evening hours.

Other uses: Toilet cleaner, fly repellent, and animal feed.

▪ PANCHAGAVYA

Ingredients (10 L preparation)

2.5 kg cow manure

250 g pure cow ghee

1 L milk

1 L curd

1 L honey

½ kg jaggery

2 L cow urine

6 bananas

Panchagavya has been used in traditional Indian rituals as well as fertilizers and pesticides since ancient times. It is a mixture of five products from the cow. The treatment is used in Ayurvedic medicine and has religious significance for Hindus.

Method of Preparation: Mix the cow manure with ghee and allow it to ferment for three days. Follow with addition of other the ingredients and mix them thoroughly. The entire mixture is stored in an earthen pot. Stir the mixture twice a day and keep it in a cool place. Close the mouth of the pot using a cloth. On nineteenth day, the preparation will be ready for use. Panchagavya can be stored for three months. When it gets solidified, dilute it by adding water.

Application: The mixture can be used by diluting 3 L of *Panchagavya* in 100 L of water. *Panchagavya* purifies crops; it can be used as a growth promoter and bio-pesticide.

▪ VERMIWASH

Vermiwash units can be set up either in barrels, buckets, or small earthen pots. The principle of this method is important. The procedure explained here is for setting up of a 250 L barrel. Take an empty barrel with one side open. On the other side, make a hole made to accommodate the vertical limb of a "T" jointed tube. Approximately ½–1 inch of the tube should extend into the barrel. On one end of the horizontal limb will be a tap, while the other end is kept closed. This serves as an emergency opening to clean the "T" jointed tube if it gets clogged. The entire unit is set up on a short pedestal made of few bricks to facilitate easy collection of the vermiwash. Keeping the tap open, a 25 cm layer of broken bricks or pebbles should be placed inside. A 25 cm layer of coarse sand then follows the layer of bricks. Water will be made to flow through these layers, enabling the basic filter unit. On top of this layer is placed a 30–45 cm layer of moistened loamy soil. Introduce about 50 surface (epigeic) and 50 sub-surface (anecic) earthworms. Cattle manure and hay will be placed on top of the soil layer and gently moistened.

The tap is kept open for the next 15 days. Water is added every day to keep the unit moist. On the sixteenth day, close the tap, and top of the unit, place a metal container or perforated mud pot at the base as a sprinkler. Pour 5 L of water (the volume of water taken out is $^1/_{15}$ the size of the main container) into this container and allow it to gradually sprinkle on the barrel overnight. This water percolates through the compost and the burrows of the earthworms, where it gets collected at the base. Open the tap of the unit the next morning and collect the vermiwash. Close the tap and refill the suspended pot with 5 liters of water that evening to collect again the following morning. Manure and hay may be replaced periodically based on need.

Application: The entire set up may be emptied and reset between 10–12 months of use. Vermiwash is diluted with water (10%) before spraying. This is found quite effective on several plants. If necessary, vermiwash may be mixed with cow's urine and diluted (1 L of vermiwash, 1 L of cow's urine, and 8 L of water) and sprayed on plants to function as a foliar spray and pesticide.

Case Study: The Effects of Organic and Chemical Agriculture on Macro and Microorganism Populations in Soil

Summary: To understand the soil health under continuous cultivation after using organic and chemical inputs, a survey was conducted on Navdanya farm areas where farmers practiced both chemical and organic inputs under different crops for at least more than five years. The effect on most important crops growing in Uttarakhand (i.e., wheat, potato, garlic, mustard, chickpea, chili, and pumpkin) were taken into consideration. The results clearly suggested a significant decline in the most important soil enzyme activities like dehydrogenase, esterase, acid, and alkaline phosphatase under chemical farming as compared to organic farming.

The microbial population—especially fungi, bacteria, actinomycetes, azotobacter, and Nitrosomonas—were significantly higher under organic

Table 6: List of micro and macroorganisms found in the soil at Navdanya farm

NAME	TYPE OF ORGANISM
Vascular Arbuscular Mycorrhizae (VAM)	
Glomus constrictum	Fungi
Glomus indica	Fungi
Gigaspora nigra	Fungi
Aculospora	Fungi
Sclerocystis rubiformis	Fungi
Phosphatase and phytase-producing microorganisms	
Aspergillus flavus	Fungi
Chitomium globosum	Fungi
Curvularia lunata	Fungi
Paecilomyces variotii	Fungi
Nitrogen fixers	
Rhizobium	Bacteria
Azatobactor	Bacteria
Soil engineers	
Earthworms	Annelida
Termites	Arthropod
Dung beetles	Arthropod
Predators	
Ants	Arthropod
Tiger beetles	Arthropod
Spiders	Arthropod
Centipede	Arthropod
Shredders	
Millipede	Arthropod

farming areas than chemical farming. There was a reduction in organic matter content of the soil under all the crops growing in chemical farming whereas increase in organic matter content under organic farming soil varies between 26–99%, although no significant changes in soil pH and EC was observed under different farming practices but a significantly higher total N and available K content was observed under organic farming practice. In general, Zn, Cu, and Fe content were significantly higher under organic farming in all the crops tested. The results clearly showed that organic farming has a great role in maintaining excellent soil health and nutrient content in the soil.

Background: A survey work has been done at Uttarakhand (Navdanya farm surrounding areas) to understand the biological soil health in organic and chemical input growing areas. In general, between 8–20 years of continuous practice was considered for sampling. The soil samples were collected from the fields of seven different crops growing under absolutely organic farming, chemical farming, and non-input conditions. The soil samples collected from bunds (barren soils) was considered as absolute control and at least four farmer's field was selected for each type of cultivation under each crop. In general, the parameters were considered as dehydrogenase, esterase, acid phosphatase, alkaline phosphatase, population of fungi, bacteria, Actinomycetes, Nitrosomonas, Azotobacter, pH, EC, Organic Carbon, N, P, K, Zn, Fe, Cu, and Mn.

Results
Beneficial Enzymes

Dehydrogenase: The dehydrogenase activity indicates the activity of bacteria and actinomycetes in the soils under different growing conditions. The dehydrogenase activity under organic, chemical, and no input conditions of seven different crops studied are presented in the following table.

Table 7: Dehydrogenase activity (pkat/g) under different crops and farming practices

CROPS	CONTROL*	NO INPUT	CHEMICAL FARMING	ORGANIC FARMING
Wheat	0.79	1.55	1.52	2.35
Potato	0.80	1.48	1.43	1.79
Garlic	0.80	1.16	1.05	1.49
Mustard	0.79	1.39	1.13	3.16
Chickpea	0.79	1.00	0.80	1.45
Chili	0.80	1.47	1.31	2.34
Pumpkin	0.78	0.92	0.71	1.28
LSD (p=0.05)	0.15	0.23	0.31	0.38

*barren land, no crops

The results clearly indicate that there was no significant difference in dehydrogenase activity in absolute control soil where no plants were growing. The dehydrogenase activity varies due to the farming practice and the crops under cultivation. The improvement in dehydrogenase activity, irrespective of crops, was much higher under organic than chemical farming (Table 7). The much higher dehydrogenase activity (300%) was observed under mustard crop and the least improvement was noticed under pumpkin (64.1%).

In general, organic farming results show 39–127% improvement in dehydrogenase activity when compared with chemical farming soils under the same crops in a similar soil condition.

The negative impact of dehydrogenase activity (2–23%) was observed when compared with no input soil with chemical farming practices, clearly indicating the adverse effects of chemical farming under different crops. When practiced, the plant contribution and soil contribution of dehydrogenase activity, it was found that there was great variation among the crops. The soil contribution was found to be much higher, in general, than plant contribution (Table 7). The overall results showed 64.2% activities of dehydrogenase contributed by soil and 35.8% were contributing by plants. In general, 18% decline in dehydrogenase activity was observed when chemical farming was practiced as compared to no input, which also clearly indicated that chemical farming has an adverse effect on soil dehydrogenase activity.

The results also showed that organic farming promotes dehydrogenase activity by 43% as compared to the crops growing under no input land. The most negative effects toward dehydrogenase activity under chemical farming were noticed on pumpkins, followed by chickpeas, and mustard.

Esterase: Esterase activity indicates the activity of fungi, bacteria, and actinomycetes in the soil under study. In general, a 2–8.7-fold improvement in esterase activity was noticed under different crop rhizosphere due to organic farming practice, which was higher under wheat, followed by mustard (Table 8).

Table 8: Esterase activity (EU x 10⁻³) under different crops and farming practices

CROPS	CONTROL*	NO INPUT	CHEMICAL FARMING	ORGANIC FARMING
Wheat	2.4	6.7	6.4	23.3
Potato	2.6	7.3	7.2	17.9
Garlic	2.8	9.7	9.4	22.6
Mustard	4.1	12.8	12.4	36.9
Chickpea	3.8	10.3	9.4	14.1
Chili	4.2	7.8	6.9	20.7
Pumpkin	3.2	5.9	5.8	12.7
LSD (p=0.05)	0.9	1.3	1.9	2.1

*barren land, no crops

Although there was little difference in esterase activity under control soil of different crops, it was very clear from the result that there was consistently higher esterase activity (28–56%) in organic farming soil compared to chemical farming. Chemical farming resulted up to 12% decline in activity as compared to no input land.

In general, an 8.4% decline in esterase activity, irrespective of crops, was noticed under chemical farming compared to no input agriculture. The decline in activity was much higher in chickpeas, followed by chilis, and wheat.

A comparison of plant and soil contribution towards esterase activity was made and it was found that 60.4% esterase activity was contributed by plants whereas soil contribution was only 39.6%. In general, more plant contribution was noticed under garlic and less was noticed under pumpkin whereas more soil contribution was noticed in chili crops.

Acid phosphatase: Acid phosphatase is mainly contributed by the plants and microorganisms in soil. Phosphatase enzymes help in the hydrolysis of the C-O-P ester bond of organic phosphorus in plant-available inorganic P in phosphate form. The activity of acid phosphatase under different input as well as seven crops (Table 9).

Table 9: Acid phosphatase (EU x 10^{-3}) under different crops and farming practice

CROPS	CONTROL*	NO INPUT	CHEMICAL FARMING	ORGANIC FARMING
Wheat	0.6	1.8	4.0	4.2
Potato	0.8	2.8	3.2	3.9
Garlic	0.9	3.4	3.4	4.0
Mustard	0.8	2.6	2.5	4.3
Chickpea	0.7	2.5	3.2	4.3
Chilii	0.8	2.9	2.8	4.4
Pumpkin	0.8	2.7	3.3	4.2
LSD (p=0.05)	0.3	0.7	0.9	0.8

*barren land, no crops

There were no differences in acid phosphatase activity under no crop condition of different crops land. However, the results showed more influence of acid phosphatase in organic farming where a 3–6-fold improvement in activities was noticed as compared to absolute control. The maximum improvement was obtained in wheat followed by chickpea. In general, 38.7% more acid phosphatase activity was found in organic farming than chemical farming (Table 9), where at least under two different crops (mustard and chili) the activities decline than no input land. Except mustard, garlic, and chili, all other crops had higher acid phosphatase activity under chemical farming as compared to no input crops.

Alkaline phosphatase: Alkaline phosphatase is only contributing by microorganisms present in the soil. They are also equally effective in breaking down the C-O-P ester bond to bring phosphorus into phosphate form for plant availability. In general, organic farming results 25–100% improvement in alkaline phosphatase activity as compared to soil of absolute control. Potato crop growing areas showed more improvement in alkaline phosphatase activity followed by garlic.

Table 9: Alkaline phosphatase (EU x 10^{-3}) under different crops and farming practice

CROPS	CONTROL*	NO INPUT	CHEMICAL FARMING	ORGANIC FARMING
Wheat	0.6	1.0	0.9	1.1
Potato	0.7	1.1	0.9	1.4
Garlic	0.8	1.2	1.0	1.5
Mustard	1.0	1.0	1.0	1.4
Chickpea	0.9	0.9	0.8	1.2
Chili	0.8	1.0	0.9	1.4
Pumpkin	0.9	1.3	1.1	1.5
LSD (p=0.05)	0.3	0.4	0.4	0.7

*barren land, no crops

The results (Fig. 9) clearly showed that there was hardly any difference in alkaline phosphatase activity under no crop (control) land but up to a 18% decline in alkaline phosphatase activity over no input land was observed under chemical farming where a 10-40% improvement in activity was noticed when farmers are practicing organic farming. The result showed tremendous contribution of organic farming on alkaline phosphatase activity. In general, the alkaline phosphatase activity under chemical farming was 73.4% less than organic farming irrespective of the crops cultivated.

Except mustard, all other crops showed negative impact of alkaline phosphatase activities under chemical farming (Fig. 10). The decline in alkaline phosphatase activity under potato was more (18.2%) followed by garlic (16.7%) and pumpkin (15.4%) while under chemical farming, which indicate an adverse effect on soil health due to chemical farming practice.

Biological parameters

Fungi population: The fungi population on different crops was increased over control soil between 6 and 36-fold when organic farming was practiced, which was much less under chemical farming (Fig. 14). Except mustard, all other crops showed decline in fungal population under chemical farming than no input cultivation. The mustard field showed there was a 59.7% improvement in the population under chemical farming which was enhanced to 14-47% further under organic farming.

Table 10: Fungi (CFU x 10 3/g) population under different crops and farming practice

CROPS	CONTROL*	NO INPUT	CHEMICAL FARMING	ORGANIC FARMING
Wheat	5.5	22	20.0	66.5
Potato	3.5	7	6.0	120.0
Garlic	7.0	20	19.5	94.0
Mustard	3.0	7.2	11.5	111.0
Chickpea	6.5	18.7	8.0	180.0
Chili	8.5	20	14.0	160.0
Pumpkin	7.0	14.1	12.0	52.0
LSD (p=0.05)	4.2	7.3	6.5	11.1

*barren land, no crops

The more increase in fungal population was observed when mustard was grown in organic farming followed by potato. In general, a 90% reduction in fungal population was observed under chemical farming as compared to organic farming growing plants although hardly any difference in population was observed under control soil. The most affected crops due to chemical farming seems to be potato and chickpea. It was noticed that plant contribution for fungal population development was much higher than soil contribution. The reduction in population due to chemical farming varies between 2.5–49.7% under different crops than no input agriculture. However, up to 16-fold improvement in fungal population was noticed due to organic farming practice when compared with no input crops (Fig. 15).

Bacteria population: Organic farming enhances bacteria population between 1.8-6.2-fold under different crops (Table 11), which was 78% more build up than chemical farming. The higher build up was noticed under pumpkin followed by chili and wheat. The plant contribution towards bacteria population was noticed between 42–66%. The reduction due to chemical farming over no input crop was between 5–33% which was more under mustard followed by garlic. In general, a 50–241% increase in bacteria population was observed under organic farming over no input land (Fig. 13).

Table 11: Bacteria (CFU x 105/g) population under different crops and farming practice

CROPS	CONTROL*	NO INPUT	CHEMICAL FARMING	ORGANIC FARMING
Wheat	2.5	4.4	4.0	15.0
Potato	3.0	8.4	8.0	12.0
Garlic	4.5	10.4	7.0	26.0
Mustard	3.5	6.0	4.0	10.0
Chickpea	5.0	9.3	7.0	14.0
Chili	2.0	5.8	5.5	12.5
Pumpkin	4.0	8.8	8.0	29.0
LSD (p=0.05)	1.7	1.9	1.8	3.5

*barren land, no crops

The adverse effect of chemical farming in bacterial population was obvious and it was more alarming especially under mustard, chickpea, and garlic (Fig. 17). The population build up under organic farming was found to be very effective under wheat followed by pumpkin among the seven crops compared.

Actinomycetes population: Organic farming builds up 47–483% more actinomycetes population under seven crops tested (Table 12). The results showed more plant contribution under mustard (52.9%) to build up actinomycetes population.

Table 12: Actinomycetes (CFU x 104/g) population under different crops
and farming practice

CROPS	CONTROL*	NO INPUT	CHEMICAL FARMING	ORGANIC FARMING
Wheat	34	43	39	50
Potato	39	45	40	67
Garlic	22	24	21	56
Mustard	17	26	24	85
Chickpea	18	23	22	105
Chili	20	24	24	45
Pumpkin	25	30	28	70
LSD (p=0.05)	7.1	11.2	8.7	13.5

*barren land, no crops

However, garlic showed (9.1%) least plant contribution towards buildup of the organisms. The reduction in activity due to chemical farming was between 0–13%, which was more under garlic and least under chili (Fig. 18).

The results showed 93% more buildup of actinomycetes population under organic farming as compared to chemical farming (Fig. 19). The maximum response to organic farming was observed under chickpea (356.5%) followed by mustard (226.9%). The least effect due to organic farming was observed under wheat (16.3%). It was very clear from the results that organic farming has definite edge over chemical farming and no input land to build up different organism's population under the rhizosphere of different crops grown under this region.

Azotobacter population: *Azotobacter* is a free-living nitrogen fixer that can fix nitrogen from the atmosphere without any outside help. In our study, their population was tremendously improved (up to 10-fold) due to organic farming practice under different crops (Table 13). It was more under mustard followed by potato and pumpkin. Although there was no significant difference in the barren land soil used for cultivation of seven crops tested.

Table 13: *Azotobacter* (CFU x 102/g) population under different crops and farming practice

CROPS	CONTROL*	NO INPUT	CHEMICAL FARMING	ORGANIC FARMING
Wheat	0.5	1.4	1.5	3.0
Potato	0.4	1.0	1.0	4.0
Garlic	0.5	0.9	0.6	1.0
Mustard	0.5	0.5	0.1	5.5
Chickpea	0.3	0.4	0.1	2.0
Chili	0.4	0.6	0.5	3.0
Pumpkin	0.5	0.8	0.7	5.0
LSD (p=0.05)	0.3	0.4	0.5	0.8

*barren land, no crops

In most of the crops showed an adverse effect on *Azotobacter* population when chemical farming was practiced as compared to no input condition (Fig. 18).

The result suggested that organic farming may boost nitrogen fixer population in the soil where almost all the crops showed 1–10-fold increase in population except garlic where organic farming resulted only an 11.1% improvement in the Azotobacter population (Fig. 18). Except under mustard all other crop shows significant plant contribution (25–64.3%) to build up Azotobacter population in the soil. The highest contribution was found under wheat followed by potato.

Nitrosomonas population: Nitrosomonas helps in transformation of nitrogen in plant available form, which was much higher (75–354%) under organic farming as compared to the chemical farming (-24-102%). Organic farming under potato resulted more buildup of population followed by wheat and pumpkin (Table 14). The results showed up to 54% influence on buildup of the Nitrosomonas population under potato followed by garlic (33%) whereas wheat crop showed least influence on buildup of Nitrosomonas population. Under wheat there was 36% reduction in Nitrosomonas population when chemical farming was practiced. The least reduction (-1.5%) was observed under potato followed by pumpkin (-9.4%). Organic farming enhanced Nitrosomonas population between 36 and 160% than no input land (Fig. 19) which was more under wheat and least under chickpea.

Table 14: Nitrosomonas (CFU g⁻¹) population under different crops and farming practice

CROPS	CONTROL*	NO INPUT	CHEMICAL FARMING	ORGANIC FARMING
Wheat	29	30	22	78
Potato	31	67	66	141
Garlic	28	42	35	60
Mustard	29	39	35	57
Chickpea	30	39	33	53
Chili	28	35	24	49
Pumpkin	30	35	32	75
LSD (p=0.05)	5.1	7.3	6.9	13.0

*barren land, no crops

The results (Fig. 21) suggested enhancing nitrifying bacterial population; organic farming has a great role.

Physio-chemical parameters

Organic matter: The buildup of organic matter was much higher under different crops when organic farming was continuously practiced. The more build up was observed under mustard and garlic (Table 10). Plant contributed more on organic matter build up under wheat (33.3%), garlic (28.6%) and chickpea (23.1%) while least contribution was noticed under chili 5.2% and potato 7.2%. In general, chemical farming resulted in reduction of organic matter build up by -14% under different crops, than no input land. The results showed 29-99% buildup of organic matter over no input land due to organic farming practiced for a long time under different crops.

Table 15: Organic matter (%) content under different crops and farming practice

CROPS	CONTROL*	NO INPUT	CHEMICAL FARMING	ORGANIC FARMING
Wheat	0.80	1.20	1.14	1.67
Potato	0.80	0.86	0.74	1.27
Garlic	0.85	1.19	1.17	2.21
Mustard	1.12	1.35	1.34	2.68
Chickpea	0.90	1.17	1.12	1.47
Chili	0.92	0.97	0.95	1.62
Pumpkin	0.85	0.93	0.85	1.29
LSD (p=0.05)	0.11	0.18	0.15	0.21

*barren land, no crops

pH: There was slight decline in soil pH (1-5%) due to organic farming than barren land under different crops (Table 16). The more reduction was observed under potato and garlic where 0.4-unit reduction in pH was noticed. The reduction in pH due to crop cultivation (no input condition) was noticed between 2–4.4%.

Table 16: pH of different crops and farming practice

CROPS	CONTROL*	NO INPUT	CHEMICAL FARMING	ORGANIC FARMING
Wheat	7.1	7	7.3	7.0
Potato	7.2	6.9	7.2	6.8
Garlic	7.3	7	7.6	6.9
Mustard	7.1	6.9	7.2	6.8
Chickpea	6.9	6.8	6.8	6.7
Chili	7.2	7.1	7.3	7.0
Pumpkin	7.0	6.9	7.1	6.9
LSD (p=0.05)	0.1	0.1	0.1	0.1

*barren land, no crops

In general, -1.4 to 4.1 changes in pH was noticed under chemical farming while slight improvement in pH was also noticed (up to 0.3 units) due to practice of organic farming. The results suggested pH does not change much both under chemical and organic farming practice.

EC: There was hardly any major change in electrical conductivity of soil due to chemical or organic farming under different crops tested (Table 17). All crops under chemical farming resulted decline in EC between 5–52%, which was more under garlic followed by mustard. The reduction of EC up to 15% was also noticed when plants were grown under no input conditions where reduction was more under wheat and mustard. In general, there was slight increase in EC under potato and garlic where organic farming was practiced.

Table 17: Electrical conductivity (EC) (dS/m) under different crops and farming practice

CROPS	CONTROL*	NO INPUT	CHEMICAL FARMING	ORGANIC FARMING
Wheat	0.20	0.17	0.16	0.11
Potato	0.18	0.17	0.15	0.27
Garlic	0.21	0.21	0.10	0.22
Mustard	0.13	0.11	0.08	0.07
Chickpea	0.12	0.11	0.08	0.09
Chili	0.18	0.17	0.17	0.16
Pumpkin	0.19	0.17	0.13	0.11
LSD (p=0.05)	0.3	0.2	0.4	0.4

*barren land, no crops

Total Nitrogen: The total N content in soil under organic farming of seven different crops tested was varies between 44–147% (Table 18), which was more under garlic followed by mustard. Except potato and pumpkin, there was no change in total N under chemical farming when compared with no input soil. It decline in total N content between 7–22% was noticed when mustard and potato was grown under chemical input.

Table 18: Percentage total Nitrogen (N) under different crops and farming practice

CROPS	CONTROL*	NO INPUT	CHEMICAL FARMING	ORGANIC FARMING
Wheat	0.08	0.11	0.11	0.16
Potato	0.08	0.09	0.07	0.13
Garlic	0.09	0.11	0.11	0.22
Mustard	0.11	0.14	0.13	0.26
Chickpea	0.09	0.11	0.11	0.14
Chili	0.09	0.10	0.10	0.16
Pumpkin	0.09	0.09	0.09	0.13
LSD (p=0.05)	0.03	0.02	0.04	0.05

*barren land, no crops

The present result suggested that to build up of N status in the soil, organic farming has major role than chemical farming or no input soil. In general, 21–100% buildup of N content was observed under different crops were regularly organic farming was practiced. The more build up (100%) over no input land was noticed under garlic followed by mustard (85.7%) and chili (60%).

Available P: Except for two crops (mustard and chickpea) organic farming enhances available P content up to 63% over no input soil. In general, very poor performance of plant contribution was noticed (5–17%) to build up available P under different crops. An erratic result was observed on available P status under chemical farming due to non-uniformity of application under different farmers field condition but more available P build up under chemical farming was notice under potato followed by chili while chickpea showed no change in available P status both under chemical and organic farming (Table 19).

Table 19: Available Phosphorus (P) (mg/kg) under different crops and farming practice

CROPS	CONTROL*	NO INPUT	CHEMICAL FARMING	ORGANIC FARMING
Wheat	21.1	24.3	28.2	33.6
Potato	25.5	30	71.8	43.7

Garlic	29.2	35.1	43.9	44.2
Mustard	28.7	34.6	46.1	30.0
Chickpea	24.0	25.4	25.3	24.9
Chili	26.0	28.7	64.3	46.9
Pumpkin	27.2	31.3	35.0	38.7
LSD (p=0.05)	1.7	2.1	3.5	3.2

*barren land, no crop

In general, sharp improvement of available P status was observed both under chemical and organic farming when compared with no input crop. The effect was more under potato and chili (Fig. 26).

Available K: Although negative impact on available K status due to chemical farming, in general, was noticed but organic farming enhances available K status under all the crops tested crops between 14–84%. The more positive effect on organic farming over no input soil was notice under garlic (84.4%) followed by chili (20.5%). Except potato and pumpkin, all other crops growing in chemical farming showed negative buildup of available K which was maximum under garlic (-22.2%). The results (Table 20) also showed least plant contribution to build up available K status in the soil. The results clearly showed garlic builds more available K status in the soil when organic farming was practiced.

Table 20: Available Potassium (K) (mg/kg) under different crops and farming practice

CROPS	CONTROL*	NO INPUT	CHEMICAL FARMING	ORGANIC FARMING
Wheat	124.3	115.6	106.6	137.3
Potato	110.9	120.3	141.7	141.4
Garlic	108.8	95	73.9	175.2
Mustard	112.0	118.7	117.7	135.5
Chickpea	110.0	115	114.0	132.2
Chili	108.0	102.6	100.5	123.6
Pumpkin	105.0	120.8	142.3	140.5
LSD (p=0.05)				

*barren land, no crops

Zinc content: Zn plays an important role in different plant metabolism processes like development of cell wall, respiration, photosynthesis, enzyme activity, and other biochemical functions. The available Zn content under different crops grown under various farming system was presented as Table 21. The results clearly showed that there was variation in available Zn under different crops, which was more under pumpkin and least under chili. No input soil, as compared to no crop land, resulted declining in Zn concentration between 2.9–6.5%. That was maximum under potato and minimum under gram. Chemical farming was influencing Zn deficiency under different crops by reducing available Zn between 15.9–37.8% while organic farming helps to restore the Zn availability in soil. The increase in availability varies between 1.3–14.3% under different crops where at least five years organic farming was practiced. The buildup was more under mustard and pumpkin while less under wheat and potato.

Table 21: Available Zinc (Zn) content (mg kg^{-1}) under different crops

CROPS	CONTROL*	NO INPUT	CHEMICAL FARMING	ORGANIC FARMING
Wheat	0.76	0.73	0.61	0.77
Potato	0.77	0.72	0.53	0.78
Garlic	0.66	0.64	0.41	0.73
Mustard	0.84	0.80	0.66	0.96
Chickpea	1.03	1.00	0.85	1.06
Chili	0.69	0.66	0.58	0.75
Pumpkin	1.29	1.21	0.96	1.41
LSD (p=0.05)	0.18	0.26	0.19	0.25

*barren land, no crops

Copper content: Cu has a role in controlling plant pathogens, which ultimately influence the yield of crops. In general, Cu content varies between 0.32–0.85 mg kg^{-1} under different soils of Navdanya farm areas (Table 22). Under no input soils there was declining in concentration between 1.4–13.2%, which was more under mustard and least under pumpkin. Chemical farming reducing Cu concentration farther between 4.2–21.3%, that can be restored by last five years continuous organic farming where improvement in Cu concentration was noticed up to 9.4%.

Table 22: Available Copper (Cu) (mg kg^{-1}) under different crops

CROPS	CONTROL*	NO INPUT	CHEMICAL FARMING	ORGANIC FARMING
Wheat	0.37	0.36	0.35	0.37
Potato	0.32	0.29	0.28	0.35
Garlic	0.59	0.57	0.55	0.60
Mustard	0.38	0.33	0.32	0.38
Chickpea	0.61	0.58	0.48	0.62
Chili	0.85	0.81	0.71	0.86
Pumpkin	0.71	0.70	0.68	0.73
LSD (p=0.05)	0.19	0.17	0.15	0.21

*barren land, no crops

When compared with no input soil, there was an improvement between 3–21% of available Cu under different crops due to organic farming. On the other hand, chemical farming reduces the Cu availability in the soil between 3–12%. The results clearly showed that the ill effect of chemical farming can be nullified by the practice of organic farming.

Manganese content: Manganese has a great role in crop physiology. It also supports the movement of iron in the plant and helps in the formation of chlorophyll. Manganese influences auxin levels in plants and high concentration of manganese favor the breakdown of indole acetic acid (IAA). The available Mn content under barren land of crops growing areas was varied between 2.16–4.66 mg kg^{-1} (Table 23).

Table 23: Available Manganese (Mn) (mg kg⁻¹) content under different crops

CROPS	CONTROL*	NO INPUT	CHEMICAL FARMING	ORGANIC FARMING
Wheat	2.16	2.01	1.78	2.26
Potato	4.07	3.98	3.48	4.66
Garlic	4.57	4.50	4.38	4.76
Mustard	2.26	2.11	2.02	2.32
Gram	4.66	4.21	4.05	4.76
Chili	3.81	3.76	3.65	3.85
Pumpkin	2.39	2.30	2.11	2.42
LSD (p=0.05)	0.83	0.72	0.98	0.77

*barren land, no crops

Due to uptake by different crops under no input treatment, the Mn content was declining between 1.3 and 9.7% with maximum under gram and minimum under chili. Chemical farming introduce further decline in Mn concentration from 4.2%–17.6% over control. There was 1–14.5% improvement in available Mn concentration due to organic farming practice under different crops, which was more under potato.

Iron content: Iron helps both as a structural component and as a co-factor for enzymatic reactions. The available iron content under barren land in Navdanya farming areas was 4.21–8.94 mgkg⁻¹ (Table 24).

Table 24: Available Iron (Fe) (mg kg⁻¹) content under different crops

CROPS	CONTROL*	NO INPUT	CHEMICAL FARMING	ORGANIC FARMING
Wheat	8.94	8.10	7.87	8.95
Potato	7.98	7.77	7.20	8.00
Garlic	7.85	7.80	7.33	7.90
Mustard	4.21	4.00	3.91	4.20
Chickpea	6.85	6.69	6.55	6.91
Chilli	7.99	7.82	7.49	8.01
Pumpkin	5.73	5.66	5.12	5.80
LSD (p=0.05)	1.01	1.23	0.98	1.19

*barren land, no crops

The available Fe content was reduced between 0.6–9.4% where crops were growing without any input, but the concentration was decreased between 4.3–12.0% due to practice of chemical farming for a longer period. However, continuous organic farming can maintain the available Fe concentration (Fig. 32) in the soil under different crops. In general, the results clearly showed that organic farming has a great role to maintain micronutrient concentration in the soil.

3.2 *Rebuilding Soil Health*

With the growing concern for sustainable development, research efforts have been focused on conservation farming, including the use of biofertilizers, organic farming, and combined protective-productive systems. The chemicalized agriculture systems are highly inefficient from an overall energy point of view, as five–ten units of energy inputs are required to produce a single unit of food energy as output (Steinhart and Steinhart 1974).

The input of fertilizers, particularly in low rainfall regions, exposes the crop to high risk. With the increased costs of petroleum and naphtha bound external inputs like nitrogenous fertilizers, the concepts of organic and conservation farming has come to stay. A sustainable approach aims to provide a means for reducing the susceptibility of soils to erosion and lowering energy-based inputs (Bethlenfalvay and Linderman 1992; Peoples and Craswell 1992). Appropriate technologies are also sought to be developed to integrate the production of crops and woody species simultaneously from the same piece of land in a sustainable manner.

Management of soils under such systems is a subject of great interest. Based on scientific evidence, the beneficial aspects of biofertilizers in agroecosystem in terms of soil fertility, nutrient cycling, soil conservation, and soil physical properties are well recognized to ensure a healthy soil plant system. This concept of sustenance of productivity is dependent on the unity and interdependence of a healthy plant-soil system in the face of natural and culturable stresses, which depend on the soundness of the interface between plant and soil: the rhizosphere.

In this era of greed and the adoption of unsustainable chemicalized farming systems, Navdanya's agroecological farm has adhered to the principle of sustainability by taking care of the water, soil, and plant components of the ecosystem. This practice has resulted in the change of an inert soil system under the former eucalyptus plantation to a living and thriving soil that is teeming with life. The quantitative improvements in soil parameters with the adoption of organic farming have been observed and analyzed in this study of Navdanya's organic farm.

The result of the study thus indicates that adopting traditional practices can enhance the fertility of the soil. This adoption will help in the enhanced agricultural output and will result in the sustained availability of natural resources. This will not only minimize the biotic pressure on agroecosystems but will also ensure long term development of the local economy.

Also in continuance, Navdanya has done a study on the changes in the percentange of organic matter in soil over a period of time. The soil samples were collected from the Navdanya organic farm, a chemical farm, and barren soil. The results showed that there was an increase in organic matter content in the soil compared to chemical farms in organic farming systems.

3.3 *Indicators of Soil Health and Ratio of Fungal to Bacterial Biomass*

By examining the structure of the soil food web in a range of soils, all grassland and most agricultural soils have ratios of total fungal to total bacterial biomass less than one (F/B<1). Another way to interpret this is that the bacterial biomass is greater than the fungal biomass. In the most productive agricultural systems, however, the ratio of total fungal to total bacterial biomass equals one (F/B=1), or the biomass of fungi and bacteria is even. When agricultural soils become fungal-dominated, productivity will be reduced. In most cases, liming and mixing of the soil (plowing) is needed to return the system to bacteria-dominated soil.

All conifer forest soils are fungal dominated, and the ratio in all forest soils in which seedling regeneration occurs is above ten. In general, productive forest soils have ratios greater than 100. This means that fungal biomass strongly outweighs the bacterial biomass in forest soils. If forest soils lose this fungal dominance, it is not possible to re-establish seedlings. When forest soil becomes bacterial-dominated, conifer seedlings are incapable of being re-established.

The ratio of total fungal to total bacterial biomass has been related to ecosystem productivity, but the amount of active bacteria and fungi are measures that also indicate soil health. For different soils, vegetation, and climates, the density of bacteria or fungi indicates the past degradation of the soil. For the most productive soils, bacterial numbers should be greater than one million for all agricultural soils, preferably closer to 100 million.

Biomass of Total Fungi

Fungal biomass is extremely important in all soils as a means of retaining nutrients that plants need in the upper layers of the soil (i.e., in the root zone). Without these organisms to take up nutrients and either retain those nutrients in their biomass or to sequester those nutrients in soil organic matter, nutrients would wash through the soil and into ground or surface water. Plants would suffer from lack of nutrient cycling into forms that the roots can take up if these nutrients aren't first immobilized in the soil through the action of fungi or bacteria.

When only fungi are present, the soil will become more acidic from secondary metabolites produced by fungi. Aggregates are larger in fungal-dominated soils than in bacterial-dominated soils, and the major form of N is ammonium since fungi do not nitrify N. These conditions are more beneficial for certain shrubs, and most trees. Total fungal biomass varies depending on soil type, vegetation, organic matter levels, recent pesticide use, soil disturbance, and a variety of other factors, many of which have not been researched completely. However, for normal grassland soils, total fungal biomass levels are usually around 50–500 meters per gram of soil. For agricultural soils, fungal biomass is around 1–50 meters per gram soil, while for forest soils, fungal biomass is between 1000 meters to 60 km per gram of soil. More work is necessary to establish what the optimal fungal biomass value should be for each type of crop, soil, organic matter, climate, etc. Very little information is available for tropical systems, but that small amount of data indicates that temperate systems perform very differently from tropical soils.

The average diameter of hyphae in most soils is about 2.5 micrometers, indicating typical mixtures of Zygomycetes, Ascomycete, and Basidiomycetes species. On occasion, the average diameter may be greater than 2.5 micrometers, indicating a greater than normal component of Basidiomycete hyphae, while on other occasions, the average diameter of hyphae may be less than 2.5 micrometers, indicating a change in species composition to a greater proportion of lower fungi. Actinomycetes are not usually differentiated from fungi, since actinomycetes are hyphal in morphology and are rarely of significant biomass. In some agricultural soils, this narrow diameter "hyphae" are of considerable importance, as demonstrated by Dr. A. Van Bruggan.

Number of Total Bacteria

Just as fungi are the most important players in retaining nutrients in forest soil, bacteria are the most important players in agricultural and grassland soils. Bacteria retain nutrients first in their biomass, and second, in their metabolic byproducts. In soil in which only bacteria are inoculated, the soil will become more alkaline, have small aggregates, and generally will have nitrate/nitrite as the dominant form of N. These conditions are beneficial for grasses and row crop plants.

Numbers of total bacteria generally remain the same regardless of the soil type or vegetation. Total bacterial numbers range between 1 million and 100 million per gram soil in agricultural soils and between 10 million and 1,000 million in forest soils. Bacterial numbers can be above 100 million in decomposing logs, in anaerobic soils, in soil amended with sewage sludge or in soil with high amounts of comported material. In some instances, following pesticide treatment, bacterial numbers can fall below 1 million, and this has been correlated with signs of severe nitrogen deficiency in plants. Bacterial numbers can drop to extremely low levels, below 100,000 per gram of soil, in degraded soils where nutrient retention is a problem.

Nematode Numbers and Community Structure

There are four major types of nematodes, which include bacterial-feeding, fungal- feeding, root-feeding, and predatory nematodes. All nematodes are predators and to some extent reflect the availability of their prey groups. However, other organisms prey upon these nematodes and nematode numbers can also reflect the balance between the availability of nematode prey, as well as feeding by nematode predators.

Both bacterial-feeding and fungal-feeding nematodes mineralize N from their prey groups. Bacterial-feeding nematodes are more important in bacterial-dominated soils (agriculture and grassland systems), while fungal-feeding nematodes are more important in fungal dominated soils (conifer and most deciduous forests). Between 70–80% of the nitrogen in rapidly growing trees has been shown to come from interactions between nematode predators and their prey. Between 30–50% of the N in crop plants appears to come from the interactions of bacterial-feeding nematodes and bacteria. Thus, the presence and numbers of bacterial- and fungal-feeding nematodes is extremely important for productive soils.

Vesicular-Arbuscular Mycorrhizal (VAM) Fungi Colonization

Vesicular-arbuscular mycorrhizal (VAM) fungi are critically important for all crop plants, except species of the Brassica family (e.g., mustards, kale). A number of researchers have shown that the lack of VAM inoculum, or the lack of the appropriate inoculum, can result in poor plant growth, weak competition with other plants, or the inability to reproduce or survive under certain extreme conditions. However, most crop fields have adequate VAM spores present, especially if crop residue is placed back into the field. Only in a few situations where soil degradation has been severe, such as with intensive pesticide use, fumigation, or intense fertilizer amendment, will VAM inoculum become so low that plant growth will be in jeopardy.

In restoration studies, the lack of appropriate inoculum is more likely to be a problem than in other situations where sources of appropriate VAM spores are near-by. Thus, the presence of at least one to five spores per gram of soil is adequate for most crop fields. When the number of spores falls below one per gram, then addition of compost containing high numbers of VAM spores (for example from an alfalfa field, or other legume), or inoculation of VAM spores from a commercial source generally results in positive effects.

At least 12% of the root system of grasses, (i.e., most crop plants) should be colonized by VAM in order to obtain the minimum required benefits from this symbiotic relationship. Colonization upwards of 40% is usually seen in healthy soils. VAM colonization can limit root-feeding nematode attack of root systems if the nematode burden is not too high. A great deal knowledge of the relationship between plant species, VAM species, and soil type, including fertility, is needed in order to fully predict the optimal relationship between crop plant, VAM species, and soil.

Disruption of Soil Fertility

The interactions between soil organisms form a web of life, just like the web that biologists study above ground. Soil biology is understudied compared to the above ground, yet it is important for the health of gardens, pastures, lawns, shrublands, and forests. If garden soil is healthy, there will be high numbers of bacteria and bacteria-feeding organisms. If the soil has received heavy treatments of pesticides, chemical fertilizers, soil fungicides or fumigants that kill these organisms, the small critters die, or the balance between the pathogens and beneficial organisms is upset, allowing the opportunist, disease-causing organisms to become problems.

Two measures of ecosystem processes are the ratio of fungal to bacterial biomass (Ingham and Horton, 1987) and the Maturity Index for nematodes. Both appear to be useful predictors of ecosystem health, although they must be properly interpreted given the succession stage being examined. For example, recently disturbed systems have nematode community structures skewed towards opportunistic species and genera, while the less opportunistic, more K-selected species of nematodes return as the time since disturbance increases. Thus, healthier soils tend to have more mature nematode community structures. However, as systems mature, nutrients tend to be more sequestered in soil biomass and organic matter, and thus the Maturity Index reflects an optimal, intermediate disturbance period in which greatest ecosystem productivity is likely to occur.

Much work is still required at the bacterial and fungal species level. While the species of protozoa and nematodes have been researched in soils of this area of the west, publication of much of this information has yet to occur. Updates will be required as this information becomes available.

Overuse of chemical fertilizers and pesticides have effects on soil organisms that are similar to over-using antibiotics. When we consider human use of antibiotics, these chemicals seemed a panacea at first because they could control disease. But with continued use, resistant organisms developed, and other organisms that compete with the disease-causing organisms were lost. We found that antibiotics couldn't be used willy-nilly, that they must be used only when necessary, and that some effort must be made to replace the normal human-digestive system bacteria killed by the antibiotics.

3.4 *Biodiverse Organic Farming to Conserve and Regenerate Water*

Most of the water pollution is because of chemical pollutants from industrial farming and livestock.

According to a UN report on the state of the world's water, "Over the last few decades, the water crisis has deepened on a planetary scale. 75% of the water use is now for chemical intensive, water intensive agriculture which also leaves the water polluted with nitrates and pesticides. More than 5 billion people could suffer water shortages by 2050 due to climate change, increased demand and polluted supplies."

By 2050, the report predicts that between 4.8 billion and 5.7 billion people will live in areas that are water-scarce for at least one month each year, up from 3.6 billion today, while the number of people at risk of floods will increase to 1.2 billion from 1.6 billion.

The FAO has also identified non sustainable industrial agriculture as the biggest contributor to the water crisis.

> **SDG Target 6.3:** "By 2030, improve water quality by reducing pollution, eliminating dumping and minimizing release of hazardous chemicals and materials, halving the proportion of untreated wastewater and substantially increasing recycling and safe reuse globally." (United Nations, 2016)

As industrial agriculture spreads, intensive irrigation spreads. In recent decades, the area under irrigation has more than doubled from 139 million hectares (Mha) in 1961 to 320 (Mha) in 2012 (FAO 2014).

The use of chemical pesticides and fertilizers continues to degrade water quality, making clean drinking scarcer. Chemical pollution raises the cost to society and is an externality not taken into account in the calculus of industrial agriculture.

Nitrate from agriculture is the most common chemical contaminant in the world's groundwater aquifers (WWAP 2013). A nationwide study in the United States estimated that farm nitrogen pollution costs Americans in the range of $59–$340 billion a year (Sobota *et al.*, 2015). The estimated annual cost of pollution by agricultural nitrogen is $35–$230 billion per year (Grinsven *et al.*, 2013). Many of these costs are associated with damages to aquatic ecosystems, deteriorating water quality, and human health impacts.

Intensive crops, livestock, and aquaculture are the main agricultural pollution sources.

As crops and animals production are separated and industrialized, the destruction of water becomes an invisible externality. And the environmental externalities of industrial agriculture are now outstripping the agricultural economy.

The total number of livestock has more than tripled from 7.3 billion units in 1970 to 24.2 billion units in 2011 (FAO 2016a). Aquaculture has grown more than twenty-fold since the 1980s, especially inland-fed aquaculture and particularly in Asia (FAO 2016b). Intensive livestock and aquaculture use antibiotics, growth hormones, and vaccines which travel from farms to ecosystems that supply us drinking water, leading to a new class of agricultural pollutants of our water supply over the last 20 years (Boxall 2012).

While feeding the world is the repeated argument used to promote water-destroying intensive systems, the monoculture, one-dimensional "yield" ignores biodiversity-based productivity and the contribution of biodiverse systems to regeneration and sustainable use of water.

Navdanya's report "Chemeenkettu" documents how the traditional shrimp and rice systems produce more food and higher net incomes for farmers than industrial aquaculture. If one further adds the costs of water pollution and ecosystem destruction by industrial shrimp farming along India's coast, the costs are higher than the benefits.

In an ecological agriculture, plants and animals are integrated and mutually support each other. Integrated, agroecological systems recycle nutrients and water, creating zero-waste systems. Ecologically, the cow has been central to Indian civilization. The integration of livestock in farming has been the secret behind India's centuries-old sustainable agriculture systems. Farm animals sustain our soils by providing soil fertility. They sustain the agricultural economy with renewable energy. As K. M. Munshi, India's first minister of agriculture after Independence, and a dear friend of my late parents, wrote: "The Mother Cow and Nandi are not worshipped in vain. They are the primeval agents who enrich the soil—nature's great land transformers—supply organic matter, which, after treatment, becomes nutrient matter of the greatest importance. In India, tradition, religious sentiment and economic needs have tried to maintain a cattle population large enough to maintain the cycle, only if we know it."

Like our seeds, India's animal breeds were bred for diversity—diversity of breeds and functions. The best cattle breeds of the world have been bred in India: the Sahiwal, Red Sindhi, Rathi, Tharparkar, Hariana, Ongole, Kankrej, and Gir. Indian breeds are multi-taskers. Both the female and male offspring have value. The cow provided nutrition through dairy, and the bullocks provided energy for transport and farm operations, and this sophisticated breeding was done by Indigenous experts.

Just as farmers' breeding of seeds and crop diversity have been ignored by industrial crop breeding, the genetic diversity of livestock with multiple uses has been ignored by the industrial animal breeding "factories", which have reduced cows and their progeny to milk and meat machines.

The industrial model breeds uniformity and one-dimensionality; it breeds standardization. Indigenous breeds in India use 29% of the organic matter provided to them compared to only 9% in US industrial farms. Indian cattle use 22% of the energy, compared to only 7% in the US.

Traditionally, cows and farm animals have used organic matter (like straw) while the grain goes to human consumption. The Green Revolution varieties deprived animals of their food. Most grain from industrial crop production is now used as animal feed, depriving humans of food. A new competition has been created between food for animals and food for humans: 75% of corn grown in India is for animal feed. In addition, we imported 500,000 tons of corn in 2016.

Yet, the highly efficient, sustainable indigenous food system, based on the multiple uses of crops and cattle, has been dismantled in the name of "efficiency" and "productivity". Integration has been replaced by fragmentation and separation. A forced one-way competition has replaced dynamic complementarity. Cyclical and circular processes—based on mutuality and the law of return—have been replaced by linearity, violence, and exploitation. India's multidimensional, multifunctional systems have been replaced by single commodity output systems using high inputs. The sacred cow has thus been reduced to a milk machine. As Shanti George observes: "The trouble is that when dairy planners look at the cow, they just see her udder; though there is much more to her. They equate cattle only with milk, and do not consider other livestock produce—draught power, dung for fertilizer and fuel, hides, skins, horn and hooves."

In the industrial-exploitative paradigm of the cow as a milk machine, our super-efficient and resilient Indian breeds are declared (quantitatively) inefficient, sans qualitative assessment. The pure indigenous breeds are replaced by homogenized hybrids of the Zebu cow, with foreign branded strains like Jersey, Holstein, Friesian, Red Dane, and Brown Swiss, supposedly to improve the Zebu's dairy "productivity."

Other contributions of farm animals are forgotten in the mechanistic reductionism paradigm. Just when we need our farm animals to play an important role in meeting the UN Sustainable Development Goals to which India is committed, we are destroying our animal wealth, and with it, the ecological and economic contributions they make. For the first time in the history of Indian agriculture, the male calves have been declared useless, which has led to the explosion of slaughterhouses and meat and beef exports. The livestock policy—made as part of the World Bank-driven structural adjustment policies—to promote the meat industry states, "religious sentiments against cattle slaughter seem to spill over also on buffaloes and prevent the utilisation of a large number of surplus male calves."

Animals on a farm sustain the soil and lives and livelihoods of small farmers. In terms of soil fertility, the slaughtered farm animals would have provided an abundance of nitrogen, phosphorous, potassium (NPK), for which we pay the "fertilizer" industry. We are not just exporting our animal wealth. We are exporting our soil and water. We are trading away our future.

We don't need to violently extract the last drop of milk from a cow and the last kilogram of a commodity from farms and crops.

We have to get rid of violence from agriculture, which is causing the disappearance of biodiversity, creating scarcity, and causing harm to plants and animals, including humans. We have to bring back care and compassion in our food and farming systems.

Agriculture based on compassion and the Law of Return provides enough for all beings, and more and better food and nutrition for humans too. The future of agriculture is conserving biodiversity, cultivating compassion. Nonsustainable industrial agriculture is the biggest contributor to India's water emergency (Vandana Shiva, *Water wars, Violence of the Green Revolution*).

The NITI Aayog, on June 17, 2018, released the results of a study warning that India is facing its 'worst' water crisis in history and that demand for potable water will outstrip supply by 2030 if steps are not taken. Nearly 600 million Indians faced high to extreme water stress and about

2,00,000 people died every year due to inadequate access to safe water. Twenty-one cities, including Delhi, Bengaluru, Chennai and Hyderabad will run out of groundwater by 2020, affecting 100 million people, the study noted. If matters are to continue, there will be a 6% loss in the country's Gross Domestic Product (GDP) by 2050, the report says.

Industrial agriculture pollutes water and destroys the water holding capacity of the soil. Hence, it requires more external inputs of irrigation. While contributing to climate change, it also makes agriculture more water vulnerable.

Not only has water-wasteful chemical agriculture mined groundwater, it has also mined soil fertility and contributed, in great part, to climate change. Chemical fertilizers destroy the living processes of the soil and make soils more vulnerable to drought. Chemical fertilizers also produce nitrogen oxide, a greenhouse gas that is 300 times more potent than carbon dioxide.

The solution for the climate, food, and water crises is the same: biodiversity-based organic farming systems. Biodiverse ecological farms address the climate crisis by reducing emissions of greenhouse gases, such as nitrogen oxide, and absorbing carbon dioxide in plants and in the soil. Biodiversity and soils are the most effective carbon sinks. They also help adapt to climate change and drought by increasing soil organic matter, which improves soil's moisture-holding capacity and hence provides drought-proofing of our agriculture.

Biodiverse organic farms increase food security by increasing the resilience and reducing the climate vulnerability of farming systems. They also enhance food security because they have higher production of food and nutrition per acre than Green Revolution monocultures, which measure the yield of one commodity, not the total food output, nor the nutritional quality of food. Biodiverse organic systems also address the water crisis as production based on water prudent crops like millets reduces water demand, due to the fact that organic systems use ten times less water than chemical systems.

By transforming the soil into a water reservoir through increasing its organic matter content, biodiverse organic systems reduce irrigation dependence and help conserve water in agriculture.

Maximizing biodiversity and organic matter production thus simultaneously increases climate resilience, food security, and water security. Many different processes lead to movements and phase changes in water.

While industrial agriculture has contributed significantly to the water crisis by disrupting the water cycle, biodiversity and organic farming address the water crisis at three levels by repairing the water cycle and regenerating water systems:

1. Biodiversity of water prudent crops reduces water demand, thus reducing withdrawals and the mining of surface and groundwater
2. Biodiversity of trees, perennials, and cover crops on-farm reduce run-off and increase infiltration, regenerating and renewing surface and groundwater
3. Biodiversity and organic farming increase the water holding capacity of the soil, making the soil a water reservoir, decreasing demand for external irrigation inputs, and increasing resilience to drought

What crops we grow and how we grow them determines whether our food system contributes to the water crisis or regenerates water systems.

Cropping patterns and water-demanding crops and varieties are primary factors contributing to the water crisis faced by communities across the world. Green revolution varieties that have been bred for chemicals also demand more water. Crops such as Green Revolution varieties of rice use as much as 3,000 to 3,500 liters for production of 1 kg of grain. This is the reason the groundwater level in Punjab has been declining rapidly.

The demands for water beyond the sustainable limits of renewability for Green Revolution monocultures of rice and wheat, has led to Punjab losing 109 cubic kilometers of water from its Indus River plain aquifer between August 2002 and October 2008 and the decline of the water table at a rate of one foot per year averaged over the Indian states of Rajasthan, Punjab, and Haryana.

Food and water are the basic needs of any society. A food system that increases the production of a few commodities by destroying the water resources without which food production is impossible is unsustainable in the most fundamental way.

In Northwest India, over-dependence on groundwater beyond sustainable level use has resulted in a significant decline in the groundwater table, leading to 16.2% of the total 6607 blocks being has categorized as 'Over-exploited' by the Central Water Board. It has categorized an additional 14% as either at 'critical' or 'semi- critical' stage.

In the 1980s, I was asked by N.D. Jayal, Adviser to the then Planning Commission, to look at why Maharashtra's requests for budgets to provide drinking water kept increasing, and yet the water crisis never gets solved. My research showed that the drought of 1972 was used by the World Bank to promote sugarcane cultivation, requiring intensive irrigation based on water mining through tubewells and borewells, just as the 1965-66 drought in India was used to push the Green Revolution, which has increased vulnerability to drought. The 2009 and 2015 droughts, and the climate crisis, are similarly being used to push the second Green Revolution with GMO seeds and patents on seeds. This will deepen Indian agriculture's vulnerability to drought.

Marathwada lies in the rain shadow of the Western Ghats and receives an average of 600-700 mm of rainfall. Given the hard rock bed of the Deccan Trap, only 10% of this water goes into the ground to recharge wells. Sugarcane requires 1,200 mm of water, which is 20 times more than the annual recharge. When 20 times more water is withdrawn from the ground than available, a water famine is inevitable, even when the rainfall is normal.

More than 300,000 farmers have committed suicide in India since 1995—most of them in the Bt cotton areas. Marathwada and Vidarbha account for 75% of farmer suicides in Maharashtra. Between January and December 2015, 3,228 farmers committed suicide in Maharashtra, including 1,536 in Vidarbha and 1,454 in Marathwada. The Chair of the Maharashtra Task Force on agrarian distress, Kishor Tiwari, has called Bt Cotton a "killer crop" which should be banned.

Sugarcane and Bt cotton have displaced crops like *jowar* (sorghum). Not only do indigenous crops like *jowar* use less water, they also increase the water-holding capacity of soil by producing large quantities of organic matter which, when returned to the soil, increases soil's fertility and water-holding capacity.

Biodiversity of native seeds and organic farming is the answer to the water crisis, drought, and climate change, to farmers' suicides and agrarian distress. They are also the answer to hunger and malnutrition. Everything is connected. The excessive groundwater extractions for industrial farming exceed the Earth's capacity to replenish its water resources by at least 160 billion cubic meters each year. Industrial agriculture accounts for up to 80% of water withdrawals in some regions.

Large quantities of water are used for farming worldwide, and some environmentalists argue this has contributed to the global water crisis. According to PeopleandPlanet.net, over two-thirds of the freshwater used by humans annually worldwide is used for crop irrigation. In Africa, for example, the Nile River loses 90% of its water for irrigation purposes before it reaches the Mediterranean Sea. In Asia, which contains two-thirds of the world's irrigated land, 85% of available water is used for irrigation. And in California, 80% of the water withdrawn for state water projects is used for agriculture. According to a report released by the environmental research and advocacy group Pacific Institute, the remaining 20% is used for residential, commercial, institutional, and industrial use.

But crops and agriculture do not have to be water intensive. 70% of the traditional rice varieties we have conserved at Navdanya do not require intensive irrigation. Crops such as millets use only 250 mm of water while they give far more nutrition. Hybrids require more water than native varieties.

Water Requirement of Different Crops

The amount of water required by a crop in its whole production period is called water requirement. The amount of water required by crops varies considerably.

Table 25: Water requirement of different crops

CROP	WATER REQUIREMENT (MM)	CROP	WATER REQUIREMENT (MM)
Rice	900–2,500	Chili	500
Wheat	450–650	Sunflower	350–500
Sorghum	450–650	Castor	500
Maize	500–800	Bean	300–500
Sugarcane	1,500–2,500	Cabbage	380–500
Groundnut	500–700	Pea	350–500
Cotton	700–1,300	Banana	1,200–2,200
Soybean	450–700	Citrus	900–1,200
Tobacco	400–600	Pineapple	700–1,000
Tomato	600–800	Gingelly	350–400
Potato	500–700	Ragi	400–450
Onion	350–550	Grape	500–1,200

Rice, wheat, and sugarcane constitute about 90% of India's crop production and these are the most water consuming crops. Green Revolution crops consume as much as 3,500 liters of water for a kilogram of grain produced.

Trees, Perennials, and Cover Crops Reduce Runoff, Increase Infiltration, and Regenerate Surface and Groundwater

Industrial agriculture has disrupted the water cycle at every step. Greenhouse gases from industrial agriculture have disrupted the climate system and impacted precipitation, leading to climate extremes, either excessive rains and floods or extended droughts.

Water that evaporates from the Earth returns as precipitation. When water vapor is condensed, it falls to the Earth's surface. Most precipitation falls on the Earth's surface as rain.

When farms and agroecosystems are covered with a green canopy of trees and cover crops, the canopy contributes to canopy interception and enhanced evapotranspiration, feeding back into atmospheric moisture. Cover crops also break the velocity of rain, thus promoting infiltration into the soil.

When there is no soil cover, up to 40% of precipitation is lost as runoff. Industrial agriculture is based on monocultures of annual crops. It removes hedgerows, farm trees, field bunds, and crop diversity from the agroecosystems. It, therefore, increases runoff.

Where rainfall lands on the soil surface, a fraction infiltrates into the soil to replenish the soil water. When the soil has organic matter, less water runs off and more infiltrates, contributing to groundwater recharge.

When soils capture the rainfall, it avoids erosion and also contributes to drought insurance.

Organic matter and biodiverse intensive organic farming contribute to regenerating water systems by increasing the capacity of soil to retain and release water. Organic matter increases infiltration, manages soil evaporation, and increases soil moisture storage capacity of the soil.

Without the protective green cover, precipitation hits the soil directly, running off, and carrying soil with it. Higher runoff translates into lower infiltration. When higher withdrawals of water are combined with lower recharge, we have a recipe of the water emergency we face.

Biodiverse mixed farming and plant residues that cover the soil surface do not leave it exposed, thus reducing runoff, which leads to soil erosion and creation of draught in the lean season. Organic matter content increases infiltration.

According to the FAO, "Conserving fallow vegetation as a cover on the soil surface, and thus reducing evaporation, results in 4 percent more water in the soil. This is roughly equivalent to 8 mm of additional rainfall. This amount of extra water can make the difference between wilting and survival of a crop during temporary dry periods."

Organic matter improves the water conserving capacity of soil by influencing its physical conditions in several ways. Surface infiltration depends on a number of factors, including aggregation and stability, pore continuity and stability, the existence of cracks, and the soil surface condition. Increased organic matter contributes indirectly to soil porosity (via increased soil faunal activity). Fresh organic matter stimulates the activity of macrofauna such as earthworms, which create burrows lined with the glue-like secretion from their bodies and are intermittently filled with worm castings.

The proportion of rainwater that infiltrates into the soil depends on the amount of soil cover provided. On bare soils (cover = 0 tonnes/ha) runoff and thus soil erosion is greater than when the soil is protected with mulch. Crop residues left on the soil surface lead to improved soil aggregation and porosity, and an increase in the number of macropores, and thus to greater infiltration rates.

Increased levels of organic matter and associated soil fauna lead to greater pore space with the immediate result that water infiltrates more readily and can be held in the soil (Roth 1985). The improved pore space is a consequence of the bioturbating activities of earthworms and other macro-organisms and channels left in the soil by decayed plant roots.

Return of organic matter to the soil transforms the soil into a habitat of rich biodiversity of soil organisms. Organisms like earthworms create water channels that increase the soil's water holding capacity and transform the soil into a dam.

As a USDA report states:

"Channels and aggregates formed by soil organisms improve the entry and storage of water. Organisms mix the porous and fluffy organic material with mineral matter as they move through the soil. This mixing action provides organic matter to non-burrowing fauna and creates pockets and pores for the movement and storage of water.

Fungal hyphae bind soil particles together and slime from bacteria help hold clay particles together. The water- stable aggregates formed by these processes are more resistant to erosion than individual soil particles. The aggregates increase the amount of large pore space which increases the rate of water infiltration. This reduces *run-off* and water erosion and increases soil moisture for plant growth."

Once infiltrated, the water becomes soil moisture or groundwater, thus regenerating and recharging water in the soil, ground, and surface water bodies such as ponds, lakes, springs, and streams.

The principles of agroecology promote practices that regenerate and renew water instead of depleting and polluting it. Planting mixtures and cover crops, integrating agroforestry perennial plants and trees with crop production, as well as building soil organic matter help the ecosystem absorb water, reduce run-off, and recharge groundwater aquifers.

The Rodale Institute reports that organic fields hold more water during droughts and that 15-20% more water seeps down to the aquifer under organic fields than does under conventional fields.

Organic soils use water more efficiently due to better soil structure and higher levels of humus and other organic matter compounds. D.W. Lotter and colleagues collected data over 10 years during the Rodale Farm Systems Trial. Their research showed that the organic manure and legume systems (LEG) treatments improve the soils' water-holding capacity, infiltration rate, and water capture efficiency.

Long-term scientific trials conducted by the Research Institute of Organic Agriculture in Switzerland comparing organic, biodynamic and conventional systems had similar results showing that organic systems were more resistant to erosion and better at capturing water.

When we started the Navdanya farm in 1994, flooding and soil erosion were major problems. We planted trees, grasses, and regenerated field bunds to reduce runoff and erosion. Biodiversity and organic farming have reduced the irrigation requirement by 75% and increased the water level by 70ft. In other words, we are returning more water to the earth than we are taking, and we are growing more food.

Biodiversity and organic farming offer solutions to many of the negative impacts of industrial agriculture on water systems.

When we think of water reservoirs, we think of large concrete dams. But the largest dam is soil with organic matter. When we return organic matter to the soil, we increase the capacity of the soil to hold more moisture. Soil moisture is the most reliable drought and climate insurance. Research across the world shows that organic farming systems improve soil structure by adding more organic matter. The more the organic matter, the higher the water holding capacity of the soil.

Water Management

One consistent piece of information coming from many studies is that organic agriculture performs better than conventional agriculture in adverse weather events, such as droughts and intense rains.

Research shows that organic systems use water more efficiently due to better soil structure and higher organic matter levels, particularly humus. The open structure allows rainwater to quickly penetrate the soil, resulting in less water loss from runoff.

Humus is one of the most important components of organic matter. It stores up to 30 times its weight in water so that rain and irrigation water is not lost through leaching or evaporation. It is stored in the soil for later use by the plants.

There is a large difference in the amount of rainfall that can be captured and stored between the current SOM levels in most conventional farms and a good organic farm with reasonable levels of SOM. This is one of the reasons why organic farms do better in times of low rainfall and drought.

The soil can be the largest reservoir of water if it has good levels of SOM. This water is stored at the crop root zone to be used as needed.

Soil organic matter increases water retention in several ways. As stated before, humus, one of organic matter's significant components, stores up to 30 times its own weight in water. The other significant role is building soil with an open sponge-like structure that efficiently captures the water and stores it in the numerous pores.

Good soil needs to be able to hold lots of water, air, and nutrients. Air is essential for the roots to breathe. Plants, like animals, need oxygen. In most cases, the roots have to get the oxygen directly from contact with the air. Some wetland plants, such as water lilies, reeds, and mangroves, have special tubes that can conduct water from the surface to the roots.

However, the majority of plants need to have their roots in direct contact with air. Too much water replaces the air, and the roots suffocate, killing the plant. Air is essential for soil microbial activity. Too much water also creates anaerobic (no air) conditions that kill the beneficial microorganisms and can favor the microorganisms that cause diseases.

The soil should be like a sponge with lots of spaces (pores) that will hold both water and air. Organic matter is the key to this.

This gives a soil numerous pores of different sizes. Some sizes are better for air, and others are better for holding water. Very importantly, the open structure of the soil ensures good capture of rain and irrigation water.

Compacted soils and those with crusts on the surface have very few spaces for water to infiltrate, so much of the water from rain or irrigation either runs off the surface or is evaporated.

Organic matter contributes to creating soil aggregates, pores, and burrows by soil fauna, increasing soil porosity and, hence, water absorption. Organic matter is food for earthworms, which fertilize the soil with castings and glue-like secretion from their bodies. Earthworms are not just the alternative to big fertilizer factories; they are an alternative to large dams. Soils rich in organic matter are rich in the diverse functions and services that soil biodiversity provides.

In addition, although they do not live long and new ones replace them annually, the hyphae of actinomycetes and fungi play an important role in connecting soil particles.

The consequence of increased water infiltration, combined with a higher organic matter content, is increased soil storage of water showed that high wheat-residue levels resulted in increased storage of fallow precipitation, which subsequently produced higher sorghum grain yields. High residue levels of 8–12 tons/ha resulted in about 80–90 mm more stored soil water at planting and about 2.0 tons/ha more sorghum grain yield than no residue management.

The addition of organic matter to the soil usually increases the water holding capacity of the soil. This is because the addition of organic matter increases the number of micropores and macropores in the soil either by "gluing" soil particles together or by creating favorable living conditions for soil organisms. Research has shown that for each 1% increase in soil organic matter, the available water holding capacity in the soil increased by 3.7%. Adhesive and cohesive forces hold soil water within the soil, and an increase in pore space will lead to an increase in the water holding capacity of the soil.

As a consequence, less irrigation water is needed to irrigate the same crop. Again, sustainable water management is a fundamental part of organic production. This ranges from the use of agronomic practices such as crop rotations, the use of green manure, catch and cover crops, which are shown to reduce nutrient leaching and run-off into water bodies. The use of synthetic pesticides and fertilizers results in huge environmental costs as a consequence of water pollution by intensive agriculture. Moreover, a strong emphasis on soil structure and increased humus quality makes organic production well placed to enhance water-holding capacity, resulting in increased resilience to extreme climate events such as heavy rainfall and droughts.

Water is a basic necessity for human and ecosystem health and necessary for the long-term ecological and socio-economic resilience of our food and farming systems. As the agri-food sector bears a large share of responsibility for water consumption and contamination it must show leadership in conserving and protecting water resources.

However, current trends in the agricultural sector place significant environmental pressure on water resources. While some progress has been achieved, poor management practices continue to have a negative impact on water quality in Europe.

A study released by Cornell University Professor David Pimentel in 2005 reported that organic farming produces the same corn and soybean yields as conventional farming and uses 30 % less energy and less water.

Moreover, because organic farming systems do not use pesticides, they also yield healthier produce and do not contribute to groundwater pollution. In addition to its conservation of water, organic farming has also been praised for the economic opportunities it creates for farmers in developing countries. Those farmers have not only found an international market for their organic products, but in drought-ridden India, organic rice farmers have found that using less water is not only a necessity but is also financially practical. Indian rice farmers cited in a 2007 World Wildlife Foundation study claimed that the system of rice intensification (SRI) helped them yield more crop with less water. Organic farming practices produce positive results for farmers and consumers. One more item to think about when you're preparing your Thanksgiving feast.

Organic Soils: A Water Reservoir

The following table shows the huge potential of organic matter not only in retaining rainwater but also reducing the soil erosion which has been expedited by the extensive use of chemicals in agriculture worldwide.

This table is designed to be a rule of thumb guide. The precise amount of water stored is dependent on soil type, specific soil density, and a range of other variables and consequently the amount could be higher or lower. However, this information is sufficient to allow an understanding of the concept.

Table 26: Volume of water retained/ha (to 30 cm) in relation to soil organic matter (SOM)

0.5% SOM	80,000 liters (common level Africa, Asia)
1% SOM	160,000 liters (common level Africa, Asia)
2% SOM	320,000 liters
3% SOM	480,000 liters
4% SOM	640,000 liters (levels pre-farming)
5% SOM	800,000 liters (levels pre-farming)
6% SOM	960,000 liters (levels pre-farming)

(Adapted from Morris, 2004)

Improved Efficiency of Water Use

Research shows that organic systems use water more efficiently due to better soil structure and higher humus and other organic matter compounds (Lotter et al., 2003; Pimentel, 2005). Lotter and colleagues collected data over ten years during the Rodale Farm Systems Trial (FST). Their research showed that the organic manure and legume systems (LEG) treatments improve the soils' water-holding capacity, infiltration rate, and water capture efficiency. The LEG maize soils averaged 13% higher water content than conventional system (CNV) soils at the same crop stage and 7% higher than CNV soils in soybean plots (Lotter et al., 2003). The more porous structure of organically treated soil allows rainwater to quickly penetrate the soil, resulting in less water loss from run-off and higher levels of water capture. This was particularly evident during the two days of torrential downpours from Hurricane Floyd in September 1999, when the organic systems captured around double the water than the conventional systems captured (Lotter et al., 2003).

Long-term scientific trials conducted by the Research Institute of Organic Agriculture (FiBL) in Switzerland, a European mountain country, comparing organic, biodynamic, and conventional systems (DOK Trials) had similar results showing that organic systems were more resistant to erosion and better at capturing water.

This is consistent with many other comparison studies that show that organic systems have less soil loss due to the better soil structure and higher levels of organic matter. (Reganold et al., 1987, Reganold et al., 2001, Pimentel 2005)

"We compare the long-term effects (since 1948) of organic and conventional farming on selected properties of the same soil. The organically-farmed soil had significantly higher organic matter content, thicker topsoil depth, higher polysaccharide content, lower modulus of rupture and less soil erosion than the conventionally-farmed soil. This study indicates that, in the long term, the organic farming system was more effective than the conventional farming system in reducing soil erosion and, therefore, in maintaining soil productivity." (Reganold et al., 1987)

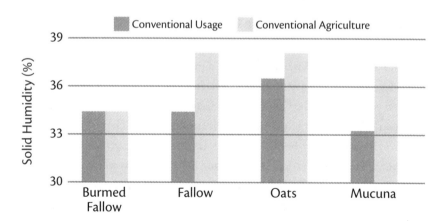

Figure 4: Soil Humidity

Figure 5: The same soil with different levels of organic matter. *Source:* Rodale Institute

The Importance of Organic Matter for Water Retention

There is a strong relationship between the levels of soil organic matter and the amount of water stored in the soil's root zone. Different soil types will hold different volumes of water when they have the same levels of organic matter due to pore spaces, specific soil density, and a range of other variables. Sandy soils, as a rule, hold less water than clay soils. For example, when referencing Fig. 5, the higher levels on the left make the soil more resistant to erosion and have higher water holding capacity. The soil on the right with low levels of organic matter holds less water, and is more prone to erosion and dispersion.

Land degradation and the water crisis are putting humanity's future at risk, and unsustainable industrial agriculture is the primary driving force for both. Regenerative organic farming regenerates both soil and water, which are the foundation of life and the economy.

Conserving and Rejuvenating Water

Destruction of water resources through water waste is one of the highest environmental costs of industrial agriculture and the Green Revolution. Large-scale intensive irrigation is not related to good agriculture or more food availability. Organic farming methods protect the agroecosystem from water runoff, evaporation, and soil erosion.

Conventional agriculture impacts the environment in many ways. It uses huge amounts of water, energy, and chemicals, often with little regard to long-term adverse effects. Irrigation systems are pumping water from reservoirs faster than they are being recharged. Herbicides and insecticides are accumulating in ground and surface waters. Chemical fertilizers are running off the fields into water systems, where they encourage damaging blooms of microorganisms. The overuse or misuse of water has not only affected the groundwater tables but also affected the quality of the soil. According to the Ministry of Water Resources estimates, during 1990–1991, about 2.46 million hectares of land in irrigated commands suffered waterlogging, and about 3.30 million hectares had been affected by salinity/alkalinity (Terra Green 2004).

It is often forgotten that 75% of agriculture is done under rain-fed conditions, and only about 25% uses irrigation. It is estimated that even if all the available water resources were developed for irrigation, about 55% of the cultivated area would still continue to be rain-fed. The Green Revolution is based on intensive irrigation and unsustainable water use, as the high-yielding varieties use much more water than Indigenous varieties.

Conservation of available soil water in agriculture is essential, as it helps plant growth. Simple techniques can be used to reduce water consumption, such as improving the efficiency of water use and reducing loss due to evaporation.

Organic farming involves many practices that protect the agroecosystem against nutrient leaching, water runoff, and soil erosion. Some of them are mentioned below.

Water Management Techniques to Reduce Water Consumption

Mulching—the application of organic or inorganic material such as plant debris, compost, etc.—in agricultural fields slows down surface runoff, improves soil moisture, reduces evaporation losses, and improves soil fertility. Crop residues are vital to the conservation of soil and water. Keeping a protective cover of vegetative residues on the soil surface is the simplest and surest way to conserve soil moisture. Vegetative residues on the soil surface improve water infiltration into the soil, reduce evaporation, and aid in maintaining organic matter. Natural mulch consists of dead leaves, twigs, fallen branches, and other plant debris accumulated on the earth's surface. Organic mulches conserve moisture and feed plants, earthworms, microbes, and other beneficial soil life. More species and tonnage of life occur below them than above the soil surface. All soil biota needs energy. They cannot collect energy directly as green plants do, but they feed on the energy released from decaying mulch, which is their preferred food source.

The experiment carried out at Navdanya Farm showed that maximum soil moisture content was recorded in the rice straw mulch field (16%) as compared to non-mulched fields (9.5%).

Mulch insulates and protects soil from drying and hard baking, caused by rapid evaporation of water from soil exposed to hot sun and winds. Mulched soils are cooler than non-mulched soils and have less fluctuation in soil temperature. Optimum soil temperatures and less moisture evaporation from the soil surface enable plants to grow evenly. Plant roots find a more favorable environment near the soil surface where air content and nutrient levels are conducive to good plant growth.

Mulches also absorb the impact of rain and irrigation water, thereby preventing erosion, soil compaction, and crusting. Mulched soils absorb water faster. Mulches prevent the splashing of mud and certain plant disease organisms onto plants and flowers during rain or overhead irrigation and help in the conservation of soil. Mulch also helps conserve moisture as it reduces 10 to 25% soil moisture loss from evaporation. Mulches help keep the soil well aerated by reducing soil compaction when raindrops hit the soil. They also reduce water runoff and soil erosion. Studies have shown that mulch also enhances the burrowing activity of some species of earthworms (e.g., Hyperiodrillus spp. and Eudrilus spp. (Lal 1976), which improves water transmission through the soil profile (Aina 1984), reduces surface crusting and runoff, and improves soil moisture storage in the root zone. Lal (1976) reports an annual saving of 3% of rainfall in water runoff from mulching in humid Western Nigeria. Roose, 1988, reports drastic reductions in runoff and erosion from a mulched pineapple field.

Improvements in soil conditions and soil water regimes to optimize runoff management techniques can support crop production. There are three main components for securing the length of the growing season to meet crop water needs:

+ Conserving water in the soil profile by allowing adequate opportunity time for rainwater to infiltrate into the soil, this is also called as *in situ* conservation of water.
+ Shaping the land surface and grading it in such a way that excess water

received during periods of high volume rainfall storms, is safely conducted to water storage reservoirs (or tanks) within the hydrologic or watershed landscape unit.

+ Augmenting groundwater recharge to ensure sustainable availability of water resources. The following methods of irrigation can reduce the soil water demand by crops.

Furrow Irrigation

Furrows are small channels which carry water down land slopes between the crop rows, allowing water to infiltrate the soil as it moves along the slope. The crop is usually grown on the ridges between the furrows. Furrow irrigation is suitable for a wide range of soil types, crops, and land slopes.

The following crops can be irrigated by furrow irrigation:

+ Row crops such as maize, sunflower, sugarcane, soybean
+ Crops that would be damaged by inundation, such as tomato, vegetables, potatoes, beans
+ Fruit trees, broadcast crops (such as wheat)

Paired Row Technique

The paired row technique is a method in which accommodating a crop grows on both sides of furrow by increasing ridge spacing; thereby a common furrow is used for irrigation of two rows. The experiments carried out by Tamil Nadu Agriculture University on green gram, black gram, groundnut, and sunflower showed that there were savings of about 20% irrigation water and 15% increase in crop yields. In the Coimbatore district farmers have adopted this technique for planting cotton, and they saved 29% of irrigation water with almost the same yield as a conventional furrow system.

Alternate Furrow System

In water scarce areas, irrigation can be applied by using alternate furrow irrigation. This involves irrigating alternate furrows rather than every furrow. Small amounts applied frequently in this way are usually better for the crop than large amounts applied after longer intervals of time.

A study conducted at Coimbatore University showed that alternate furrow saves irrigation water compared to all-furrow irrigation. The data are presented in the following table.

COUNTRY	LOCATION	YIELD ON BED (KG/HA)	YIELD ON FLAT BED (KG/HA)	VS. FLAT (%)	WATER SAVING MING
Bangladesh	Dinajpur	4,710	3,890	25	
India	Punjab	4,530	4,220	24	
	Haryana	5,290	5,010	46	
	UP	4,750	4,550	30	
Kazakhastan	Almaty	5,080	4,900	29	

Source: www.fao.org

Bed System (Raised or Flat)

The bed system depends on the intensity of rains and type of soil. Bed systems give higher yields of 54-80%. The water savings by using raised bed methods are presented in the following table:

SOWING MAT BOD	WHEAT PRODUCTIVITY (1/ACRE)	SOIL MOISTURE (%)
Ridge method (paired row)	11.54	12.0
Railed bed method	10.25	11.5
Flatbed method (conventional)	7.15	8.5

An experiment on wheat productivity (Q/acre) by using different methods of sowing was carried out at Navdanya's experimental field. The data showed maximum wheat productivity in Ridge method (paired row). The maximum soil moisture was also recorded in the Ridge method.

Runoff management and conservation of soil water by organic farming practices are based on the principles of minimizing the concentrations of run-off volume, slowing the runoff velocity, so diminishing its capacity to cause erosion. It aims to enhance surface detention storage, thus allowing the water more time to soak into the soil. Biological control measures used in organic farming practices, combined with good agronomic and soil management practices, provide better protection of the soil from raindrop impact, increase surface depression storage and infiltration capacity of soil to reduce the volume of runoff, improve soil aggregate stability to increase its resistance

to erosion, and increase the roughness of the soil surface to reduce the velocity of runoff. Mulching appears to be the effective conservation measure of
organic farming practices.

Figure 6

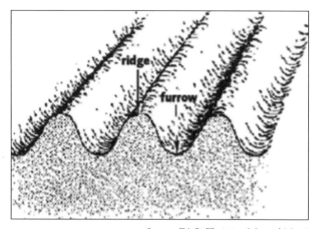

Source: FAO, Training Manual No. 5

✍ CASE STUDY ✍

Water Efficient Sugarcane Farming from South India

Suresh Desai, a sugarcane farmer in the Belgaum district of Karnataka State, South India, has developed a series of modifications to the conventional package of practices associated with sugarcane farming. According to Desai, the conventional practice of flooding the crop's root zone damaged soil aeration, reduced soil fertility, and consequently made the plants susceptible to diseases. He redesigned the irrigation channels and began by reducing their number by half. This he did by eliminating every alternate channel. Thus, for every channel that he kept, one was eliminated. The channel that was eliminated earthed up and turned into a bed of mulch in order to facilitate the retention of moisture in the soil. He discovered that the sugarcane field as a whole was now able to retain a greater amount of moisture. This method reduced water supply to the field area by 50%. The number of irrigations required also decreased.

After three months, the number of channels was reduced to two. With this method, he was able to raise four rows of plants with only one channel of water. In the conventional method, other farmers maintain four water channels for four rows of sugarcane. These modifications reduce the requirement of water in irrigated sugarcane plantations by approximately 75%.

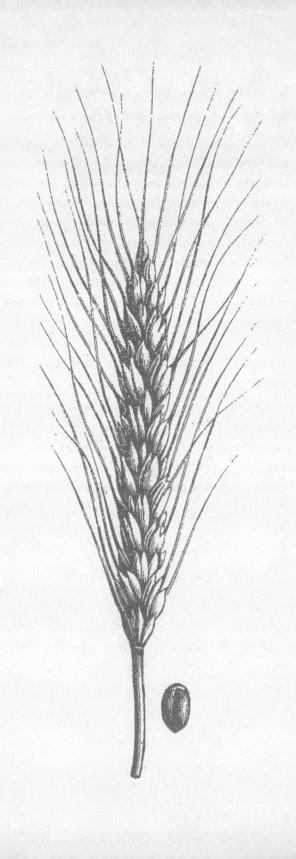

Atmospheric CO_2 levels have increased by over 2 parts per million (ppm) per year. Despite all the commitments countries made in Paris in December 2015, the levels of CO_2 increased at record levels in 2016. 3.3 ppm of CO_2 entered the atmosphere. In 2020, levels of CO_2 reached a new record of 412.5 ppm, the highest level recorded in 800,000 years.

> 1 ppm = 7.76 Gt CO_2
>
> 2pm = 15.52 Gt CO_2 (per year)

According to the World Meteorological Organization:

"Geological records show that the current levels of CO_2 correspond to an 'equilibrium' climate last observed in the mid-Pliocene (3–5 million years ago), a climate that was 2–3 °C warmer, where the Greenland and West Antarctic ice sheets melted and even some of the East Antarctic ice was lost, leading to sea levels that were 10–20 m (30 to 60ft) higher than those today."

Even if the world transitions to 100% renewable energy tomorrow, this will not stop temperatures and sea levels from rising because it will take more than 100 years for the CO_2 levels to drop. As a result of natural disasters and hazards, millions of people will be displaced, throwing the planet into chaos. Not only should we speed up the transition to renewable energy and reduce carbon emissions, but it is necessary to use methods of adaption to draw down CO_2 currently in the atmosphere.

Utilizing regenerative methods such as plant photosynthesis and biodiversity will give humanity the ability to remove excess carbon dioxide from the atmosphere and return it to the soil, where it can contribute to food, water, and climate security.

There is already an excess of carbon in the oceans, causing many problems for ecosystems and sea life. Soils are the most logical sink for carbon, with estimates that they store over 2,700 gigatons of CO_2. This is more than the atmosphere (848 Gt) and biomass (575 Gt) combined (Lal 2008).

Agriculture can have a major role as fertilizers, manufacturing methods, chemical transportation, and farm inputs influence whether it is a problem or a solution. Depending on externalities, greenhouse gas emissions in the agricultural sector can range from 30–50%, but there is the potential to reduce them significantly. Degradation of soil and desertification have exacerbated the depletion of soil carbon, with estimates indicating that agricultural soils have lost 50-70% of their soil organic carbon (SOC).

Restoring the organic carbon pool can be accomplished through long rotations, catch crops, cover crops, green manures, legumes, organic agriculture, compost, organic mulches, biochar, perennials, agroforestry, agroecological diversity, and livestock on pasture using sustainable grazing systems.

The following are examples of organic farming systems that sequester more CO_2 than they emit and their potential for total carbon sequestration when applied globally across all agricultural lands.

+ Manured organic plots in the Farming Systems Trial at the Rodale Institute sequestered CO_2 at a rate of 3,596.6 kg of CO_2 per hectare per year. If it were extrapolated globally across agricultural lands, it could sequester 17.5 Gt of CO_2 annually.

+ A meta-analysis of 24 comparison trials in Mediterranean climates between organic and non-organic systems found that the organic systems sequestered 3,559.9 kilograms of CO_2 per hectare per year. When the data is extrapolated globally across agricultural lands, these systems will sequester 17.4 Gt of CO_2 per year (Aguilera *et al.*, 2013).

+ The Louis Bolk Institute conducted a study to assess soil carbon sequestration at Sekem, Egypt's oldest organic farm. For the past 30 years, Sekem's management methods have been able to sequester 3,303 kgs of CO_2 per hectare each year on average. Based on these figures, global adoption of these practices has the potential to sequester 16 Gt of CO_2 into soils each year.

+ CO_2 has been proven to be sequestered in the soil at a rate of 8,220.8 kg of CO_2 per hectare each year as part of the Rodale Compost Utilization Trial. If extrapolated globally, this would sequester 40 Gt of CO_2 per year.

Pastures account for the bulk of the world's arable agricultural land (4,883,697,000 ha or 68.7%), according to the Food and Agriculture Organization (FAO, 2010). There is a growing body of published evidence suggesting that pastures may build up SOC faster than many other agricultural systems and store it deeper in the soil under good management:

> "In a region of extensive soil degradation in the southeastern United States, we evaluated soil C accumulation for 3 years across a 7-year chronosequence of three farms converted to management-intensive grazing. Here we show that these farms accumulated C at 8.0 Mg ha^{-1}yr^{-1}, increasing cation exchange and water holding capacity by 95% and 34%, respectively." (Machmuller *et al.*, 2015)

To put these figures in perspective, consider that (8.0 Mg ha^{-1}yr^{-1}) means 8,000 kg carbon is stored in the soil per hectare per year. Multiplied by 3.67, these grazing systems have captured 29,360 kg of CO_2 per hectares per year (29.36 metric tons CO_2/ha/year). If these regenerative grazing practices were implemented on the world's grazing lands (3,356,940,000 ha), they would sequester 98.5 gt CO_2/yr.

As shown in the preceding examples, various agricultural systems might store enough CO_2 to make a significant difference in our efforts to counteract climate change. It is critical that further studies be conducted to discover how and why these systems capture substantial amounts of CO_2, as well as how to scale these findings for use on a global level in order to achieve a meaningful amount of GHG reduction. The rates at which sequestration occurs may be enhanced through additional research.

The immediate goal must be to stabilize the CO_2 in the atmosphere to prevent future increases in climate-related extreme events. This should ideally be accomplished through a combination of emissions limitations and the adoption of renewable energy and energy efficiency policies. However, the Paris Agreement stipulates that this will not happen until 2030 at the soonest, indicating that the widespread use of regenerative farming practices may make a significant contribution to CO_2 stabilization and reduction prior to 2030. Governments, international organizations, industry, and climate change organizations should make the widespread adoption of these systems the highest priority.

4.1 The Climate Crisis: Transgressing Planetary Boundaries and Disrupting Ecological Cycles

The climate crisis is the most dramatic expression of human impact on planet Earth. While the Earth's climate has gone through several phases of warming and cooling throughout history, the current trend toward warming and associated climate system disruption and weather patterns are human induced. Tragically, those who have contributed the least to emissions bear the brunt of the impacts—villages in India's high Himalayas that have lost their water sources owing to glacier melt and disappearance, residents in India's Ganges basin whose crops have failed owing to drought, and coastal communities are threatened by sea level rise and intensified cyclones.

Extractive agricultural techniques that rely on fossil fuels are causing ecological processes and planetary boundaries to break down. Industrial chemical agriculture is based on external inputs of nitrogen, phosphorous, and potassium; industrial monocultures are based on globally traded commodities.

Exacerbated by a combination of factors, including habitat destruction and pollution, global biodiversity is rapidly eroding. Industrial agriculture that relies on fossil-fuel intensive, chemical-intensive monocultures uses 75% of the land yet produces only 30% of our food supply while small, biodiverse farms utilizing 25% of the area produce 70%. Monocultures, particularly those found in industry, are a major contributor to the loss and erosion of biodiversity. Land is being cleared in the Amazon and Indonesian rainforests for Roundup Ready soy and palm oil monocultures.

We used to consume 10,000 plant species. Today, just 12 globally traded commodities are cultivated. Even more, only 10% of corn and soy is consumed as food while the rest is turned into biofuels and animal feed. At this rate, if our diet's industrial agriculture and industrial food portions rise to 45%, the planet will be lifeless. Rejuvenating and regenerating the environment through ecological processes has become an essential survival imperative for humanity as well as all other species. The change from fossil fuels to living processes that are based on developing and recycling living carbon is critical to the transition.

4.2 *Regenerating the Living Carbon Cycle*

Life on Earth depends on the soil, sunlight, and seeds. Within this living economy, all of humanity's and other animals' needs are met in a sustainable manner. As Sir Albert Howard writes in the Agriculture Testament:

> "The energy for the machinery of growth is derived from the sun, the chlorophyll in the green leaf is the mechanism by which this energy is intercepted; the plant is thereby enabled to manufacture food-to synthesize carbohydrates and proteins from the water and other substances taken up by the roots and the carbon dioxide of the atmosphere. The efficiency of the green leaf is, therefore of supreme importance: it depends on the food supply of the planet, our well being, and our activities. There is no alternative source of nutriment. Without sunlight and the green leaf our industries, our trade, our possessions would soon be useless." (Howard 1940)

With the aid of the sun, seeds germinate and develop into plants that form the Earth's green covering, returning organic matter to the soil and providing humans and animals with all of their resources for food, clothing, and housing.

The primary entrance of carbon into the biosphere is through plant photosynthesis or gross primary productivity, which is the uptake of carbon from the atmosphere by plants. Carbon can be lost through plant respiration (autotrophic respiration), as a result of litter and soil organic matter decomposition (heterotrophic respiration), and as a consequence of additional losses caused by fires, drought, human activities.

The capacity to store carbon may be limited as a result of climate change, which causes ecosystems degradation. A warming planet can increase heterotrophic respiration and decomposition in the soil's organic matter content. Therefore, carbon stock may be a very useful tool until a more ecologically acceptable alternative replaces dependence on fossil fuels.

4.3 *Navdanya Climate Change Adaptation Study*

Navdanya conducted a study in four distinct ecological regions of India to assess the impact of organic farming on climate change adaptation and measure numerous characteristics, including water retention capacity, soil carbon accumulation, carbon sequestration, microbial biomass, biological activity, enzyme activities, effect on crop and cropping system, soil physical properties, and soil organic carbon stabilization and loss. In Uttarakhand, case studies were collected, comparing organic farming to conventional farming for paddy and sugarcane.

The organic matter in soil is made up of plant and animal residues, as well as other organic compounds produced by soil microorganisms during decomposition. This is continuously broken down and resynthesized by soil microbes. As a result, organic matter (soil carbon) is a transitory soil component that exists for several hours to hundreds of years. Maintaining soil carbon is critical for mitigating the effects of climate change, and organic farming is a significant way it can be done.

A higher increase in total carbon build-up due to organic farming compared to chemical farming was studied under different agroecosystems. The results showed an additional increase in soil carbon for organic agriculture irrespective of crop growth (Table 1). In general, there was greater carbon accumulation in humid agroecosystems. The additional carbon build-up was more prevalent under organic farming, which would promotes soil microbial growth, nutrient recycling, and moisture retention in the soil. It also aids in the prevention of soil erosion, especially in arid and semi-arid regions.

Table 1: Increase in carbon build up due to organic agriculture*

	ADDITIONAL INCREASE (μ g-1)	
	Range	*Mean*
Arid	49-83	62.5
Semi-arid	57-98	71.9
Sub-humid	61-101	75.5
Humid	68-102	83.0

*Average of 10 farms in each agroecosystem

Living carbon is the cycle of life that involves living seed, living soil, and the life-giving sun. Living carbon is not comparable to fossilized carbon. The disruption of the carbon cycle and climate system destabilization as a result of extracting fossil fuels (dead carbon) from the planet, burning them, and releasing uncontrollable emissions into the atmosphere, which ruptures the carbon cycle and causes climate system destabilization.

All of the coal, petroleum, and natural gas we are extracting was formed over the course of 600 million years. The dependence on fossil carbon leads to a scarcity of living carbon, which depletes food supplies for people and soil organisms. Food security is a major challenge, especially in areas of conflict and natural disasters. This scarcity leads to malnutrition and hunger, as well as soil degradation. Chemical agriculture increases capital inputs while simultaneously decreasing biodiversity, biomass, and nutrition that may be produced from the seed, soil, and sunlight. We must biologically intensify our farms and forests in terms of both biodiversity and biomass to address the problem of carbon pollution from the atmosphere.

The more biodiversity and biomass we grow, the more the plants fix atmospheric carbon and nitrogen and reduce both emissions and the stocks of pollutants on the air. Carbon is returned to the soil through plants. That is why the connection between biodiversity and climate change is an intimate connection. The more the biodiversity and biomass intensification of forests and farms, the more organic matter is available to return to the soil. This reverses the trend towards desertification, which is the primary reason for displacement and uprooting of people, which creates climate refugees.

Organic farming—working in accordance with nature—takes excess carbon dioxide from the atmosphere, and through photosynthesis, puts it back in the soil. It also increases the water holding capacity (WHC) of soil, contributing to resilience in times of more frequent droughts, floods, and other climate extremes. Organic farming has the potential of sequestering 10 gigatons of carbon dioxide, equivalent to the amount needed to be removed from the atmosphere to keep atmospheric carbon below 350 ppm and the average temperature increase lower than two degrees Celsius. We can bridge the emissions gap by utilizing the techniques of ecological agriculture now.

To repair the broken carbon cycle, it's imperative to increase the living carbon in plants and soil. Working with living carbon gives life; using fossilized carbon disrupts living processes. Dead carbon must be left underground as an ethical obligation and ecological imperative.

This is why the term "decarbonization" without qualification and distinction between living and dead carbon is scientifically and ecologically inappropriate. If we decarbonized the economy, we would have no plants, which are living carbon. We would have no life on earth which creates and is sustained by living carbon. A decarbonized planet would be a dead planet. Rather, there must be an approach to recarbonizing the world with living carbon and decarbonizing it of dead carbon.

4.4 *The Carbon Wheel*

Photosynthesis stores carbon in the soil. The goal of photosynthesis is to build up the carbon reservoir in the lithosphere. This accumulation of "carbon wealth" does not sit idle. It becomes the groundwork for terrestrial life, both below and above the soil surface. The Carbon Wheel, in other words, is a dynamic version of our lithospheric carbon pool.

A wheel is a positive sign of progress, hope, and happiness. These attributes represent the "spokes" of the Carbon Wheel. Photosynthesis constructs the Carbon Wheel when carbon dynamically moves from its atmospheric

Figure 1: The carbon wheel.

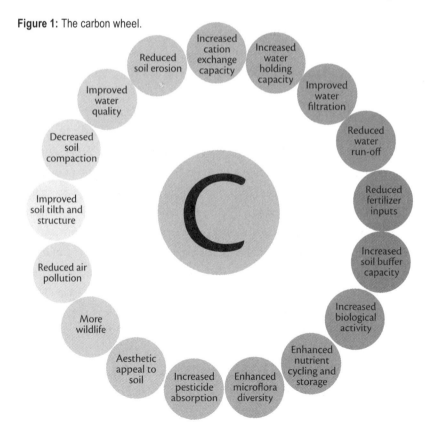

pool to all varieties of life through green vegetation, finally making its deposits in the lithosphere. Fed by photosynthesis, our environment will continue stabilizing, the biosphere will remain healthy, and our future will progress in the mode of sustainability (Reicosky 2007).

The more atmospheric carbon that enters into life through photosynthesis, the more constructive the Carbon Wheel becomes. When the phenomenon of photosynthesis is obstructed to a certain extent, the "spokes" of the Carbon Wheel break and more carbon is emitted back to the atmosphere where it causes climate change. This is what is happening in our contemporary times. When "spokes" of the Carbon Wheel are intact, and bound in the lithosphere, the wheel gets stronger, moves on in balance, and carbon contributes to more growth.

The Carbon Wheel of the lithosphere has its constructive impression on all of the biosphere, weaving life everywhere. It keeps moving within the lithosphere and it builds up the pathways to humanity's material and cultural progress. It upholds the Living Planet in balance and sustainability. The Carbon Wheel of the lithosphere keeps moving, and the cosmos goes on writing its mysterious story of evolution.

On Earth, the cosmic evolution flowers with photosynthesis. All of the existing and continuously evolving species and all of humankind are the beautiful flowerings of this evolution. Whatever we see, smell, hear, touch, and feel is all on account of photosynthesis. Whatever we conceive and cultivate within is all owing to photosynthesis. Myriad colors in nature, all varieties of life, all breathtaking ecstasies, and the infinite beauty that we witness are the lively gifts of photosynthesis. We, the humans, have evolved as custodians of the biosphere, which was an indomitable will of photosynthesis.

All evolution on Earth is a benevolence of photosynthesis, smiling on us, for we are the most wonderful beings of it. So wonderful that photosynthesis generated a unique consciousness in us, and we were evolved, as guardians of photosynthesis itself.

Our hands cannot be cruel, We cannot enslave the phenomenon that controls the climate of our own destiny. Let us awaken to the consciousness of benevolence that photosynthesis deeply ingrained in us. Let us liberate photosynthesis, give it back its full freedom, and help it prevail with its all potencies. Then we shall also prevail amidst a climate that showers its benevolence upon us to help us prevail with glory and happiness.

4.5 *Fossil Fuel Based Synthetic Fertilizers*

The last century has seen the rise of fossil fuel-based, chemical-intensive industrial agriculture. All pesticides marketed by companies like Monsanto and Syngenta are based on fossils fuels. Fossil fuel-based farming is the biggest contributor to climate change, with 40–50% of greenhouse gases coming from it (Shiva 2008).

In addition to carbon, nitrogen fertilizers produce nitrogen oxide emissions, which is a greenhouse gas that 300 times the warming potential of carbon dioxide in the atmosphere. Nitrogen fertilizers are not just destabilizing the environment, they also create dead zones in the oceans and desertify the soils. In the planetary context, the erosion of biodiversity and transgression of the nitrogen boundary are serious crises. These aspects of ecological disaster are often overlooked.

The process by which synthetic nitrogen fertilizers are manufactured is based on fossil fuels and used to make explosives and ammunition during World War II. Following WWII, when huge stockpiles of unused ammonium nitrate munitions were offered for agricultural usage, synthetic nitrogen fertilizer was promoted in agriculture. The Haber-Bosch method, which uses natural gas to artificially fix nitrogen from the atmosphere at high temperatures and manufacture ammonia, is highly energy intensive. The energy required to make one kg of nitrogen fertilizer is equivalent to two liters of diesel. In 2000, the energy used in fertilizer production was equivalent to 191 billion gallons of diesel, and it is expected to rise to 277 billion gallons by 2030. This has significant effects on climate change, but it is largely neglected. Phosphate fertilization consumes half a liter of diesel per kilogram (Shiva 2008).

Synthetic fertilizers, like other fossil fuels, harm the carbon cycle. They also disrupt the nitrogen cycle and the hydrological cycle since chemical agriculture requires 10 times as much water as organic farming to produce the same amount of food. Additionally, they pollute rivers and seas through contamination.

Since war expertise does not provide expertise about how plants work, how the soil works, and how ecological processes work, the militarized industrial agriculture model ignored the potential of biodiversity and organic farming.

But solutions can be found in the practices of agroecology. Returning organic matter to the soil builds up soil nitrogen. A recent Navdanya study shows that organic farming has increased soil nitrogen content between 44–144%, depending on the crops. Pulses fix nitrogen nonviolently in the soil rather than encouraging reliance on synthetic fertilizers generated through the use of fossil fuels at 550 degrees Celsius. Chickpea can fix up to 140 kg nitrogen per hectare, and pigeon-pea can fix up to 200 kg nitrogen per hectare. Integrating pulses in organic agriculture is the only sustainable path to food and nutritional security. This is the integration of life and the intensification of ecological processes, not the integration of power and intensification of chemicals, capital, and control. This is how ancient cultures enriched their soils.

Vegetable protein found in pulses is also an essential component of a healthy, balanced diet for humans. The "benevolent bean" is fundamental to the Mediterranean diet. India's culinary culture is based on "*dal roti*" and "*dal chawal*." *Urad, moong, masoor, chana, rajma, tur, lobia,* and *gahat* are our main staples. India was formerly the world's largest producer of pulses—our proteins are nutrient-dense and flavorful. The Green Revolution's monoculture has driven pulses out of the country, and now Bt cotton and soy are threatening to do the same. In 2014, 11.6 million hectares were planted with Bt seeds. If pulses had been grown in half this area, we would have had an additional 4 million tons of legumes available. As a result, we are losing almost 10 million tons of pulses.

The results of a study at the Navdanya farm show that organic matter has increased by up to 99%, Zinc by 14%, and Magnesium by 14% without adding external inputs. They have been generated by billions of soil microbes that are present in healthy soils. Healthy soils produce healthy plants, which may then be consumed by people. Chemical farming, on the other hand, has resulted in soil nutrient depletion, resulting in a reduction in the nutritional value of our meals.

Pulses are truly the pulse of life for the soil, for people, and the planet. In our farms, they give life to the soil by providing nitrogen. This is how ancient cultures enriched their soils. Farming did not begin with the Green Revolution and synthetic nitrogen fertilizers.

Whether Navdanya, Baranaja, or the "three sisters" planted by the first nations in North America, or the ancient Milpa system of Mexico, beans and pulses are vital to Indigenous agriculture. As Sir Albert Howard, known as the Father of Modern Agriculture, writes in *The Agriculture Testament*, comparing agriculture in the West with agriculture in India:

"Mixed crops are the rule. In this respect the cultivators of the Orient have followed nature's method as seen in the primeval forest. Mixed cropping is perhaps most universal when the cereal crop is the main constituent. Crops like millets, wheat, barley, and maize are mixed with an appropriate subsidiary pulse, sometimes a species that ripens much later than the cereal. The pigeon pea (cajanusindicus), perhaps the most important leguminous crop of the Gangetic alluvium, is grown either with millets or with maize... Leguminous plants are common. Although it was not until 1888, after a protracted controversy lasting thirty years, that Western science finally accepted as proved the important role played by pulse crops in enriching the soil, centuries of experience had taught the peasants of the east the same lesson." (Howard 1940)

Table 2: Showing effect of continuous farming on soil under organic and chemical mode

NUTRIENT	CHANGE UNDER CHEMICAL FARMING	CHANGE UNDER ORGANIC FARMING
Organic Matter	-14%	+29–99%
Total Nitrogen (N2)	-7–22%	+21–100%
Available Phosphorous (P)	0%	+63%
Available Potassium (K)	-22%	+14–84%
Zinc (Z)	-15.9–37.8%	+1.3–14.3%
Copper (Cu)	-4.2–21.3%	+9.4%
Manganese (Mn)	-4.2–17.6%	+14.5%
Iron (Fe)	-4.3–12%	+1%

From the 1990s, there has been a debate in the policy, academic, and civil society circles on the ill effects of chemical fertilizers on soil health and food security. The government of India acknowledged the problem only in 2009 when then Union Finance Minister Pranab Mukherji in Parliament during his budget speech said:

"In the context of the nation's food security, the declining response of agricultural productivity to increased fertilizer usage in the country is a matter of concern. To ensure balanced application of fertilizers, the Government intends to move towards a nutrient based subsidy regime instead of the current product pricing regime ..."

Healthy soils are the foundation for food, fuel, fiber, and even medicine. Chemical fertilizers are destroying the soil food web and the living organisms that create soil fertility, soil aggregates, and help conserve water in the soil. Industrial agriculture, therefore, creates higher vulnerability to climate change by contributing to desertification and drought, which affect food security and livelihood security.

4.6 *Biodiversity-Based Organic Farming for the Mitigation and Adaption of Climate Change*

According to the United Nations Framework Convention on Climate Change (UNFCC 2011), the world is seeing an increase in the frequency of extreme weather events such as droughts and heavy rainfall. Even if greenhouse gases were immediately eliminated from the atmosphere, it would take many decades for climate change to reverse itself. This indicates that farmers must learn to cope with increasingly severe and frequent adversities, including droughts and heavy, destructive rainfalls.

Studies show that organic farming systems are more resilient to predicted weather conditions and can produce greater yields than conventional farming systems under such circumstances (Drinkwater , Wagoner, and Sattantonio1998; Welsh, 1999; Pimentel, 2005). The Wisconsin Integrated Cropping Systems Trials found that organic production was higher during drought years and comparable to conventional production in non-drought years. (Posner et al., 2008)

There is a large difference in the amount of rainfall that can be captured and stored between the current SOM level in most traditional farms in Asia and Africa and a good organic farm with reasonable levels of SOM. This is one of the reasons why organic farms do better in times of low rainfall and drought.

In drought years, the Rodale Farming Systems Trials (FST) showed that organic farming methods produced more corn than conventional agriculture.

Corn yield rates were 6,938 and 7,235 kg per ha in the two organic systems during drought years, compared to 5,333 kg per ha in the conventional method (Pimentel, 2005). The higher yields in the dry years were attributed to organic farms' superior ability to absorb rainfall. This is owing to the greater amounts of organic carbon present in those soils, which makes them more brittle and able to hold and capture rainwater, allowing it to be used for agriculture (Rodale 2011).

When the soil is saturated, it becomes difficult to cultivate hemp organically. When mechanical weed cultivation is delayed in wet years, organic yields are 10% lower (Posner et al., 2008). Instead of tillage, using steam or vinegar to control weeds might help correct this. The higher crop yields in dry years, according to the researchers, are due to organic farms' improved ability to take up rainfall quickly. This is because the soils are more friable and able to store and capture rain since to greater amounts of organic carbon.

The Cornell University study of the Rodale Field Study revealed that conventional crops perished during drought years, while organic crops fluctuated little during drought years as a result of enhanced soil water retention capacity in the improved soil (Pimentel et al., 2005). When these yield fluctuations were averaged out, the organic crop had yields that were equivalent to or greater than those of the conventional crop.

In drought years, organic systems produced more corn than conventional systems. In those five drought years, average corn yields were greater (28% to 34%) in the two organic systems: 6938 and 7235 kg per ha in the organic animal and legume management strategies, respectively, than they were in the standard system (5,333 kg per ha) (Pimentel 2005).

Higher yields in dry years were attributed to the ability of organic farm soils to absorb more rainfall. The greater amounts of organic carbon in the soil, the more friable and able to retain and capture water. According to the authors, "This yield advantage in drought years is due to the fact that soils higher in carbon can capture more water and keep it available to crop plants." (La Salle and Hepperly, 2008)

This is extremely valuable knowledge, as the majority of the world's agricultural production systems are rainfed. The world does not have enough water to irrigate all of its agricultural areas. Nor should such an endeavor be initiated because damming the globe's waterways, extracting drinking water from all subterranean aquifers, and constructing hundreds of thousands of miles of canals would be an unprecedented environmental catastrophe. Organic farming methods are the most efficient, cost-effective, ecologically sustainable, and practical solution for ensuring consistent food production in a changing climate.

The organic food and agriculture movement is gaining in strength in spite of the monumental opposition of agrochemical industries, whose economic existence depends on synthetic fertilizers and pesticides. The movement is gathering momentum as farmers are increasingly becoming aware that industrialized chemical farming entails an ever-increasing production cost and rapidly declining soil fertility, crop yield, and livelihood security.

Sustainable food production must be based on the restoration of biodiversity, soil, and water in a localized environment. It can also conserve natural capital such as biodiversity, soil, and water while boosting nature's economy, improving farmers' livelihoods, enhancing the security of agricultural workers' jobs, and improving the quality and nutrition of our foods (Shiva, 2008).

According to a study of the US food chain, organic and chemical farming require ten calories of energy to generate one calorie of food. In addition, when organic farming was carried out, the fungal population on various plants increased by 6–36 times over control soil (2.5–49.7%). Organic farming raises bacteria populations between 1.8 and 6.2-fold under diverse crops, which is 78% greater than chemical agriculture (2.5–49.7%).

4.7 *Agrobiodiversity, Climate Resilience, and Sustainability*

Recently, Navdanya investigated the effects of crop diversity in food security and economic sustainability in five regions of Uttarakhand, two areas in Bundelkhand, and one region in Maharashtra and Rajasthan. In the study, crop loss due to untimely rainfall during both the ripening and harvesting period was observed. Results revealed a positive correlation between decreasing agrodiversity and a quantitative increase in crop loss. Increasing diversity within the species coupled with the use of traditional open-pollinated strains shows increased food and economic security against climate change-related crop damage.

According to government statistics, as a result of altered weather conditions during the *rabi* crop season, over two million tons of pulse crop production was lost. In Rajasthan, Maharashtra, Uttar Pradesh, and Dehradun and

Chakrata in Uttarakhand, wheat output was reduced by 30–70%. Only the traditional variety of lentils known as *Teen Fool Wali Masoor* could survive in pulses, whereas all other lentil types could not withstand the altered weather conditions. In Rajasthan and Lalitpur, where there was a wider range of crops, the proportion of crop failure decreased. Farmers in Maharashtra's Banda and Chakrata regions, where diversity was lower, experienced significant crop loss.

Non-industrial agriculture, on the other hand, saves up to seven times more energy and offsets 5–15% of global fossil fuel emissions through carbon sequestration in organically managed soil. Up to 4 tons of CO_2 can be trapped per hectare each year in organic soils (Shiva 2008).

When organic manure is used as a fertilizer, it improves soil fertility and has fewer negative consequences on the environment without sacrificing crop productivity, according to studies by marginal farmers in the developing world and scientists. Organic fertilizers decompose slowly, releasing nutrients gradually. This slow nutrient release helps to build carbon and nitrogen in the soil, which reduces leaching losses (Jenkinson et al., 1994). Productivity on organic farms is constant as the nutrient cycling is tighter in the agroecosystem due to organic inputs. Such is not the case with synthetic chemical inputs—a lesson that agriculture must learn again.

Manure-based farming systems have been shown to boost soil organic matter and total nitrogen content by 120% in comparison with conventional fertilizer application on farmland (Jenkinson et al., 1994; Powlson, 1994). The Rodale Institute's Farming Systems Trial (FST) is America's longest-running, side-by-side comparison of organic and chemical farming. In 1981, The FST surprised a food community that still mocked organic practices when they began to study what happens during the transition from chemical to organic agriculture. Yields of organic systems increased after an initial downturn during the first few years of the transition, and they soon returned to match or surpass conventional systems. Over time, FST became a means for comparing the long-term prospects of the two systems.

Corn and soybeans made up over 49% of all croplands in the United States. Other grains accounted for 21%, forages for 22%, and vegetables just 1.5% of the total acreage. The FST has included three fundamental farming systems over its lengthy history, each with its own set of management techniques:

+ A manure-based organic system
+ A legume-based organic system
+ A synthetic input-based conventional system

To represent farming in America more accurately today, genetically modified (GM) crops and no-till techniques have been added to the study. Results and comparisons are accordingly labeled to reflect this change.

According to Rodale Institute studies, organic yields not only match but, in some cases, exceeded conventional yields during drought years. Organic farming systems also contribute to soil organic matter rather than depleting it, making it more sustainable and energy-efficient. In comparison to organic farming, conventional systems generate 40% more greenhouse gases. It also shows that organic farming methods are more profitable than traditional farming practices.

Navdanya has been promoting and researching agriculture that conserves biodiversity, improves farmers' seed sovereignty and food sovereignty, and increases nutrition per acre, while also enhancing small farmer income, thus simultaneously tackling poverty, malnutrition, hunger, and climate change for the last three decades.

Multiple cropping in the same soil and climatic conditions is more economically beneficial than current intensive chemical farming methods involving monocultures, according to a study conducted by Navdanya in West Bengal's four districts (Deb 2004). According to the research, productivity rises when crops are combined with animals and the relative value of farm produce increases significantly with a greater diversity of crops.

A result of many years of selection in accordance with current agro-hydrological systems, inaccessible resources, and ecological fragility is biodiversity-based traditional farming systems. These circumstances culminated in the development of subsistence production methods that were sustained through the organic matter and nutrients derived from the forests.

In a study of the *Rabi* and *Kharif* seasons of 2014–15, Navdanya discovered that organic farming is significantly superior to chemical farming in nine distinct regions in five Indian states, including Maharashtra, Odisha, UP, Uttarakhand, and Rajasthan. All nine areas are distinct from one another. Within Uttarakhand, Rajasthan is an arid zone while Bundelkhand and Maharashtra are drought-prone regions, yet Odisha is a flood-prone location.

Dehradun is a valley with an elevation of approximately 500m AMSL, and Purola Valley has an elevation of 1500m AMSL. Rudraprayag and Tehri are among the state's hill districts.

The table below summarizes the findings, which show that crops produced in organic farms have outperformed those grown in chemical farms. This research is being done with 1,074 farmers from five Indian states who switched to organic in 2013 as a result of Navdanya's help. *Dehradun* Basmati rice, red paddy, wheat, corn, mustard, *tur, urad, moong, jeera,* and lentil are among the crops studied.

Table 3: Showing comparative productivity analysis of chemical vs. organic farms in Rabi 2014/15 and kharif season 2015

Area	Crops	Production of chemical farms (acre in qtl)	Production of organic farms (acre in qtl)	Average increase (%) of organic farms
Amravati, Maharashtra	Cotton	11.0	13.0	18.18
	Tur	3.8	5.3	39.47
	Wheat	11.7	11.8	0.85
Tonk, Rajasthan	Mustard	5.3	5.43	2.45
	Wheat	10.3	11.2	8.74
	Moong	4.8	5.3	10.42
	Urd	3.8	4.1	7.89
	Jeera	1.23	2.1	70.73
Lalitpur, Uttar Pradesh	Wheat	8.2	8.3	1.22
	Lentil	2.2	2.67	21.36
	Urd	1.12	2.31	106.25
	Moong	1.43	2.11	47.55
Banda, Uttar Pradesh	Tour	1.45	1.55	6.90

The average percentage increase was as high as 106.25%, with a range of 0.85% to 106.25%. Organic farms outperformed chemical farms in every test, including those under severe climate stress, which demonstrates that organic farming is superior in all circumstances, regardless of the location or crop.

4.8 *Biodiversity: A Climate Solution*

Biodiversity is our main line of defense against climate change. Diversity shields us from both climatic extremes and uncertainty. Biodiversity increases a region's resilience to climate change by enhancing the soil's capacity to resist drought, floods, and erosion.

The ability of living systems to adapt and evolve is a sign of their diversity. That is why, researchers are turning their attention to agrobiodiversity preservation and evolutionary breeding. Natural environments have been able to adapt naturally or autonomously to changing circumstances due to agricultural biodiversity in natural ecosystems. As the severity of climate change increases over time, the necessity for co-evolution for adaptation becomes more acute.

Traditionally, villages that rely on biodiversity riches have informal systems and customary rules in place to avoid that external changes go beyond natural resilience levels. *Van Panchayat* is one such example; it still exists in many parts of India today. According to the UNEP, incorporating social and ecological factors into climate change adaptation plans is critical. The time-tested, age-old solutions might need to be bolstered by contemporary formal adaptation tactics in order to confront new threats to biodiversity, taking into account the rapid rate of changes in demographic, economic, and sociocultural environments.

The in situ and ex situ conservation of crop and livestock genetic resources is important for maintaining options for future agriculture needs. In situ conservation of agricultural biodiversity is defined as the management of a diverse set of crop populations by the farmers in the ecosystem where the crop evolved. It maintains the processes of evolution and adaptation of crops to their environment. Ex situ conservation involves the conservation of species outside their natural habitat, such as in seed banks and greenhouses.

The conservation of the components of agricultural ecosystems that provide goods and services, such as natural pest control, pollination, and seed dispersal, should also be promoted. Indeed, 35% of the world's crop production is dependent on pollinators such as bees, birds, and bats.

Biodiversity increases genetic diversity, which is indispensable to cope with environmental stresses and is the cornerstone of small farmers' livelihood strategies. It is also the basis for food security as it provides alternatives to fossil fuels and chemical inputs for small-scale and ecological farms. Biodiversity is the only ecological insurance for society's future adaptation and evolution in the face of extreme weather patterns. Increasing genetic and cultural diversity in food systems and maintaining this biodiversity in the commons are vital adaptation strategies to respond to the challenges of climate change.

Monocultures, centralization, and techno-fixes represent a myopic obsession that must give way to diversity and decentralization. Biodiversity and small-scale farms go hand in hand, yet corporate-driven globalization policies that promote monocultures are pushing farmers off the land; policies that protect and expand biodiversity must be encouraged to mitigate the impact of climate change.

The resilience of ecosystems can be enhanced and the risk of damage to human and natural ecosystems could be reduced through the adoption of biodiversity- based adaptive and mitigation strategies. Mitigation is described as a human intervention to reduce greenhouse gas sources or enhance carbon sequestration, while adaptation to climate change refers to adjustments in natural or human systems in response to climatic stimuli or their effects, which moderates harm or exploits beneficial opportunities.

Mitigation and adapting to climate change are encouraged by the following examples of projects:

+ maintaining and restoring native ecosystems
+ protecting and enhancing ecosystem services
+ managing habitats for endangered species
+ creating buffer zones
+ conservation of local flora and fauna (including agricultural crops and their landraces)
+ promotion of biodiversity-based ecological farming
+ documentation of Indigenous knowledge

The wealth of biodiversity communities all around the globe derive many essential goods and services from natural ecosystems such as food, fresh water, timber, fuelwood, fiber, non-timber products, genetic materials, etc. People ate the grain, and long straw was fed to cattle which enriched the soil with their excrement, which provided food for microorganisms that fed the crop, and the cycle was unbroken until recently. The human economy clearly depends upon the services by ecosystems, carried out "for free." Natural ecosystems also perform fundamental life support services without which human civilizations would cease to thrive. Since the beginning of life on Earth, human beings developed knowledge and found ways to derive livelihoods from the bounties of nature's diversity, in wild as well as in domesticated forms. It is evident that a certain level of biodiversity is necessary to provide the material basis for human life: at one level to maintain the biosphere as a functioning system and, at another, to provide the basic materials for agriculture and other utilitarian needs.

Hunters and gathers in the beginning of civilization used thousands of plants and animals for their food, medicine, shelter and clothing. This number is coming down with so-called development. People are now dependent on very few plants for their livelihood, which created imbalance in the nature by promoting monocultures, exploitation of certain resources, and indirectly imposing pressure on Earth to fulfill the greed of humans. Diversity is the characteristic of nature and the basis of ecological stability. It is also a concept, which refers to the range of variation or differences among some set of entities. Biodiversity simply means the biological diversity, which refers to variety within the living world. The term is used commonly to describe the number, variety, or variability of living organisms. Simply stated, the entire variety of plants, animals, and all other living organisms on the Earth constitutes the biodiversity of our planet.

Biodiversity is not merely the genetic components of diverse species, but the interrelationships among the flora and fauna including: microorganisms, soil, water, ecosystems, the environment, and the cosmos as a whole. The diverse climatic and ecological zones of our country provide a congenial setting for the evolution of a wide range of ecosystems. From the tropical Western Ghats to the temperate Himalaya and from the fertile coastal regions to the cold deserts of Ladakh, India supports a strikingly diverse and rich range of biodiversity.

Many of the human activities that modify or destroy natural ecosystems may cause deterioration of ecological services whose value, in the long term, dwarfs the short-term economic benefits society gains from those activities. Fortunately, the functioning of many ecosystems could be restored if appropriate actions were taken in time. Climate change, including variability and extremes, continues to impact ecosystems sometimes beneficially, but frequently adversely on their structure and functions.

Erosion of Agro-Biodiversity

Green Revolution farming practices involving homogenization of the crop genetic base has eroded biodiversity in agro-ecosystems including plant genetic resources, livestock, beneficial insects, and soil organisms. Further, replacing Indigenous varieties with high biomass—therefore, high organic matter and a bigger contribution to the living carbon cycle—with dwarf varieties adapted to chemical fertilizers disrupted both the carbon and nitrogen cycles.

Indigenous crop varieties were most suited to providing ecological functions and services, providing for human and animal needs. Grain was eaten by people, long straw was fed to cattle, who in turn, enriched the soil with their dung. This dung was food for microorganism who in turn provided food for the crop. This cycle was completely broken during the last 60 years by the Green Revolution based on chemical monocultures of dwarf varieties, thus reducing food for animals and for the soil.

Synthetic pesticides are responsible for the decline in spider, bee, wasp, beetle, cricket, dragonfly, damselfly, earthworm species diversity and abundance as well as pesticide resistance in crop pests and pathogens of nontarget species. Rachel Carson's 1961 classic *Silent Spring* on the impact of DDT on bird eggshell thinning has prompted a wave of research that documents the role of pesticides in erosion of biodiversity.

Indigenous crops varieties can tolerate a wide range of climatic and soil conditions, whereas modern crop strains are more prone to perish as a result of minor environmental changes, such as too early or late rains. Crop landraces grown by traditional farmers continue to evolve genetically in response to human management and environmental changes. A large array of genes responsible for resistance to different pests, pathogens, and environmental conditions are found in folk crop cultivars and their wild relatives. The loss of traditional varieties threatens the genetic foundation for crop breeding and improvement. Fearing the irretrievable disappearance of valuable genes, conservationists have launched efforts to collect and save folk crop seed samples for future use in ex situ gene banks (Jackson 1995).

The Green Revolution has led to high external input based intensive agricultural systems from the traditional self-reliant agricultural system. About 7,000 plant species have been cultivated for food since agriculture was practiced by human beings. Today, however, only about 15 plant species and eight animal species supply 90% of our food. Many traits incorporated into modern crop varieties were introduced from wild relatives, improving their productivity and tolerance to pests, disease, and difficult growing conditions. Wild relatives of food crops are considered an insurance policy for the future, as they can be used to breed new varieties that can withstand changing conditions.

Agricultural modernization has destroyed the genetic base of most cultivated crops—rice, wheat, soy, and potato—by replacing them with a few contemporary types (Fujisaka 1999). Many wild strains of important food crops are at risk of disappearing. For example, one-quarter of all wild potato species are expected to go extinct in 50 years, posing a potential barrier to future plant breeders ensuring that commercial cultivars can withstand a changing climate.

It is estimated that there were over 200,000 different varieties of rice grown across 41 million hectares, with 60 million tons of rice produced each year. It's been shown that the narrow genetic base in Indian rice is a result of the HYV containing dwarfing gene from Taichung I and IR (Richharia and Govindswami 1990).

Climate Change: An Anthropogenic Threat to Biodiversity

Biodiversity is rapidly declining owing to climate change, but appropriate management of biodiversity can help to mitigate the effects of global warming. There are several scientific studies that demonstrate how climate change is already having an impact on biodiversity and will continue to do so. The Millennium Ecosystem Assessment classifies climate change as one of the major immediate drivers affecting ecosystems. Major consequences of climate change on species biodiversity include:

+ Changes in distribution pattern of the species
+ Increased vulnerability and extinction rates
+ Changes in reproduction timings
+ Changes in growing seasons for plants

Many species that are already endangered are particularly susceptible to the effects of climate change (WWF online report). The extinct golden toad and *Monteverde harlequin* frog were identified as the first casualties of climate change (Pounds, Fogden, and Campbell 1999). Any reduction or alteration in rainfall affects frog development owing to frogs' requirement of water for reproduction. Furthermore, high temperatures are closely linked to the spread of a fungal disease that contributes to amphibian population decline. The tigers' habitat in Asia's mangrove forests may be destroyed as a result of predicted sea level rises.

Crop diversity is an important safeguard against climate change in any particular farm. We've incorporated this time-tested knowhow into Navdanya's biodiversity farm in Doon Valley (Uttarakhand), which is based on biodiversity and farming in nature's ways. In the field, as I experimented with the mixed cropping system in a number of combinations of seven, nine, and twelve crops (*baranaja*), we discovered that mixed biodiverse crops consistently outperformed monocultures. They are also resistant to frost, droughts, early, late, or even minimal rainfall.

Traditionally, farmers have increased their resilience by growing more than one crop. At Navdanya's biodiversity farm in Doon valley (Uttarakhand), we have build on this ancient knowledge, farming based on nature's principles. At farm while experimenting with the mixed cropping system in several combinations of seven, nine, and 12 crops (*baranaja*), we found that mixed biodiverse crops always performed two to three times better than that of monocultures. They are also capable of tolerating the frost, drought, early, late, or even very little rains (Shiva, 2008).

Multifunctional, biodiverse farming systems and localized diversified food systems are required for food security in a changing climate. It is critical both for mitigating climate change and maintaining food security to make a quick worldwide transition to such systems. This study is a testament to our firm belief that we will survive climate chaos only if biodiversity and its nurturing conditions thrive, and that climate resilience and adaptation methods stay in the commons, not corporate hands. The evolution of nature and farmers' breeding have maintained genetic diversity, and farmer's breeding has allowed agriculture to adapt over the last 10,000 years, and it will play a significant role in adjusting farming to climate change in the decades ahead.

Industrial mechanistic and reductionist solutions supplant intimate understanding of biodiversity and ecosystems with careless technologies. Agrichemicals and genetic engineering destroys and depletes the very biodiversity, soil, air, and water that agriculture requires while also exacerbating climate change. The genetic diversity of plants, as well as the rich knowledge and practices of farming communities, are the two most essential resources for adapting agriculture to changing climatic conditions.

Crop genetic diversity is important for coping with environmental pressures, and both traditional and Indigenous knowledge systems embrace key principles of adaptation, diversity, and plurality. Public policy and investment are required to recognize and promote crop genetic diversity in order for communities to adapt to climate change.

The corporate-led "climate-resilient" gene campaign, which promotes patented seeds that will not enable small farmers to adapt to climate change, is a distracting and deceptive public relations effort by seed businesses attempting to portray themselves as climate saviors while concealing the underlying causes of climate change and genuine solutions.

Patented "techno-fix" seeds will not allow the adaptation strategies that small farmers, especially the most vulnerable poor farmers, need to cope with climate change. These proprietary technologies will ultimately concentrate corporate power, drive up costs, inhibit independent research, and further undermine the rights of farmers to save and exchange seeds. These patented solutions represent a violation of farmers' knowledge, a commons accessible to all, and people's rights to be able to develop climate adapting strategies.

How Biodiversity Makes Agriculture and Communities More Resilient to Climate Change

There are four ways in which biodiversity and seed freedom creates climate resilience and is a climate solution:

A. Farmers have bred climate resilient seeds and varieties that are contributing to resilience.

B. Diversity of crops increases the resilience of farming to climate change. If you have only one crop in a monoculture, it is more vulnerable to the changing climate. Farmers growing monocultures of commodity crops are also more vulnerable to exploitative markets.

C. Biodiversity intensification allows more carbon to be absorbed from the air and returned to the soil, thus decreasing excess carbon in the atmosphere while also increasing the resilience of soils to drought, floods, and climate change.

D. When farmers have their own renewable, regenerative seeds, they can replant after a climate disaster, which contributes to both climate resilience and economic resilience. If farmers are dependent on purchase of costly nonrenewable seeds from corporations, not only do they lose their crop, they lose their sovereignty and are forced into debt. Debt is the single biggest reason for the more than 300,000 farmers suicides in India since 1995.

Climate Change Requires Farmers' Breeding and Local Adaptive Strategies

Plant breeding plays an essential role in adapting agriculture to rapidly changing climates. Even when formal sector scientists use the most sophisticated climate models and the most advanced technologies, the reality is that they are not very good at predicting what happens at a very local level and on the ground realities. While genetic uniformity is the hallmark of commercial plant breeding, farmers and breeders, rooted in local level realities, deliberately create and maintain more heterogeneous varieties in order to withstand diverse and adverse agroecological conditions. The crop diversity developed and maintained by farming communities already plays a role in adapting agriculture to climate change and variability. Additionally, farmers adapt quickly to changing climates by shifting planting dates, choosing varieties with different growth duration, changing crop rotations, diversifying crops, and using new irrigation systems—among other strategies. Farmer-led strategies for climate change survival and adaptation must be recognized, strengthened, and protected. Farming communities must be directly involved in setting priorities and strategies for adaptation.

Farmers' knowledge and technology have never been stagnant or static. They have always skillfully responded to changing circumstances and have kept their system in a dynamic state, advancing towards higher degrees of complexity, resilience, sustainability, and security. In the process of achieving these goals, farmers have always based their livelihood systems on natural biodiversity. They unabatedly searched, selected, cultivated, bred, preserved, protected, saved, conserved, experimented with, managed, used, enriched, shared, distributed, and disseminated the germplasm, which is the living testimony of their innovations. Not only this, they also dutifully passed this germplasm on to the next generations.

Systematic modern agricultural experiments are just about a half-century old, whereas farmers' experiences are millennia old. They cannot be ignored or rejected as mere remnants of the past. Farmers' knowledge and technology are futuristic and innovative and are governed by ecological laws. They must find central place in our contemporary agricultural strategies. Recent advances in technology will be welcomed, provided they have compatibility with those evolved by farmers and rooted in local realities (Singh *et al.*, 2013).

While highly expensive, high-yielding seeds, hybrid seeds, and GMOs continue to fail. Indigenous open-pollinated, climate resilient varieties are proving to be an important option for adaptation to climate change. Evidence from farmers' fields proves that Indigenous crop varieties can withstand a wide range of climatic and soil conditions, whereas "modern" crop varieties tend to perish at small environmental variations like rains arriving too early or too late. Farmers' varieties grown by traditional farmers continue to evolve to adapt to changing environmental conditions. With the disappearance of biodiversity due to industrial monocultures, the very genetic base for crop breeding for climate resilience is irretrievably lost. Fearing the loss of valuable genes, conservationists have launched efforts to collect and save folk crop seed samples for future use in ex situ gene banks (Jackson 1995).

Navdanya's experience of working with farmers across the country reveals that climate resilient seeds with organic farming are better than "high yielding" seeds in chemical farming. A study done by Navdanya in Odisha, Bundelkhand, Uttarakhand, and Maharashtra confirms that open-pollinated Indigenous seeds are better alternatives to the hybrid, high-yielding, or GM seeds. Hundreds of farmers in Odisha were given Indigenous seeds by Navdanya after the Phailin super cyclone and they got very good yields.

As insurance against such vulnerability, Navdanya has pioneered the conservation of biodiversity in India and built a movement for the protection of small farmers through promotion of ecological farming and fair trade to ensure healthy, diverse, and safe food. Navdanya's program for promoting ecological agriculture is based on biodiversity for economic and food security. Today, as a result of Navdanya's pioneering work, many small groups and entrepreneurs have entered in the field of biodiversity conservation, organic farming, and marketing of organic food products.

Navdanya's experience of working with farmers across the country and Bhutan established that through adopting the principles of agroecology and biodiversity-based organic farming, farmers could not only increase their yields by two to three times, but *vis-à-vis* can reduce their input costs. Indigenous open-pollinated varieties are not only capable of producing more, but are also resilient to the climate. Comparative studies of 22 rice-growing systems have shown that Indigenous systems are more efficient in terms of yields, labor use, and energy use (Shiva & Pandey, 2004).

Researchers have already proven the importance of Indigenous crops and organic farming practices in coping with the changing climatic conditions. Results of our studies in the past in different agro-climatic situations confirms that even in adverse climatic conditions, biodiversity-based organic farming (higher crop diversity) is better capable to minimize the crop losses than that of monoculture-based industrial farming.

Climate resilient traits will become increasingly important in times of climate instability. Along coastal areas, farmers have evolved flood tolerant and salt tolerant varieties of rice such as *"Bhundi," "Kalambank," "Lunabakada," "Sankarchin," "Nalidhulia," "Ravana," "Seulapuni," "Dhosarakhuda."*

Crops, such as millets, have been evolved for drought tolerance, and provide food security in water scarce regions, and water scarce years. Corporations like Monsanto have taken 1,500 patents on climate resilient crops. Navdanya, the research foundation for Science, Technology and Ecology, published the list in its report, *Biopiracy of Climate Resilient Crops: Gene Giants Steal Farmers Innovation.*

Navdanya chose to protect the vanishing rice varieties of Odisha by preserving their genetic diversity using both in situ and ex situ methods as well as conducting tests on their sustainability in a variety of ecological settings, including rapid climate change and yield potentials under various soil amendments. This was useful when selecting the seeds of specific rice diversities to empower local communities to restore agriculture in disaster areas like Erasama in Odisha after the Orissa super cyclone, Nagapattinam in Tamil Nadu after the Boxing Day tsunami of 2005, and Nandigram in Bengal after the Boxing Day earthquake.

Navdanya has also given hope to the victims of the tsunami. The tsunami waves affected the agricultural lands of the farmers due to intrusion of seawater and deposition of sea land. More than 5,203 hectares of agricultural land in Nagapattinam were destroyed during the tsunami. The Navdanya team conducted a study in the affected villages to facilitate the agriculture recovery. The team distributed three saline resistant varieties of paddy—including *Bhundi, Kalambank,* and *Lunabakada*—to the farmers of the worse affected areas. These varieties of native saline resistant kharif paddy seeds were collected from Navdanya farmers in Orissa amounting to a total of 100 quintals.

Navdanya Odisha currently maintains four seed banks, three village level and one central level, where seeds of diverse rice varieties are conserved and renewed every year. Climate resilience is given importance in the village level seed banks where all available rice land races are conserved in the central seed bank. Navdanya also encourages individual cultivators to save, exchange, and increase diversities in their own fields. The village level seed banks are located in different and varied eco-climatic zones, like salt-prone, flood-prone and drought-prone areas. The central seed bank has 810 rice varieties in its accession, out of which 119 varieties are climate resilient. 33 of these are salt and flood-tolerant including 1 aromatic variety, 47 are flood-tolerant, and 39 are drought-tolerant including 3 aromatic and 2 therapeutic rice varieties. The rest 581 varieties belong to the general category. There are 56 aromatic rice varieties, of which 2 have unique and diverse aroma, 1 smelling like fried green gram and the other, like cumin seed are not available anywhere in the world. The therapeutic rices are used in old age tissue rejuvenation.

Seed exchange has been the backbone of paddy cultivation until the Green Revolution. Native paddy plants have diverse basal sheath colors, with about 9 shades of 5 colors, ranging from green, yellow, purple, violet to black. Reappearance of wild varieties is an inherent characteristic of paddy cultivation. Cultivators, hence, replace the variety with a different basal sheath color next season just to be able to distinguish it from the weeds which are then manually removed. All the Green Revolution varieties have the same basal sheath color, making it difficult to distinguish the wild weed, which is never removed. A particular variety cultivated in a given field for more than 3 years lose yield, hence, is replaced. This replacement used to be procured through seed exchange, a part of the barter system that was in place until a few decades ago. Thus the cultivators used to gain twice, a new variety and higher yield as the new variety always yielded more. The Green Revolution proponents do not contribute to this gospel truth. It has been further found out that seeds exchanged over a long distance for growing in the same type of micro-climate not only yielded much more but often even changed its potentials. Two examples will suffice to put all doubts at rest:

1. *Udasiali*, an Indigenous photosensitive *kharif* paddy variety transported over 500 kilometers from Balasore to Erasama in Jagatsingpur as part of post-1999 super cyclone disaster agricultural rehabilitation yielded on par wih *rabi*.
2. Three select Odisha salt-tolerant paddy varieties transported over a distance of over 1,500 kilometers from Balasore to Nagapattinam in Tamilnadu under the 'Seeds of Mope' Program following the 2004 tsunami yielded three times more and far better than any known high yielders. The same varieties behaved even better when cultivated in Indonesia, another 1,000 or more kilometers away, in 2006 by Professor Friedhelm Goltenboth of Hohenheim University, Germany.

Climate-Resilient Seeds to Cope with Climate Change

With the increasing events of disasters, we started conserving climate resilient seeds. We also encouraged farmers to grow and multiply native climate-resilient varieties and created "Seeds of Hope," to help disaster-affected farmers with climate-resilient seeds.

It is predicted that a 4°C increase in temperature due to climate change will reduce rice yield by 10%. Rice has been found to be quite climate resilient. Rice as a crop originally flourished in the dry climate of central Asia and later spread to wet tropical Asia, thus evolved the lowland rice varieties with better yield.

Odisha is very well known for its rice diversity; therefore, Odisha was selected to conserve and multiply climate-resilient paddy and vegetable seeds. Climate-resilient varieties conserved by Navdanya in Odisha are given below.

Salt, Flood, and Drought Tolerant varieties

Navdanya has conserved 33 salt-tolerant varieties. Odisha salt-tolerant rice landraces have caused miracles both in Nagapattinam and Indonesia (post-tsunami), and some of them, such as *Lunabakada, Kalambank, Bhundi*, and Dhala sola, have on an average produced 35–54 tillers in the SRI method.

In the last 20 years, Navdanya has conserved 54 flood-tolerant varieties in Odisha. Of these, eight varieties are extremely water-tolerant. These varieties are being conserved and multiplied at Navdanya's biodiversity conservation farm and Seed Bank in Chandipur, Balasore, and by the Navdanya member farmers in Odisha.

One of the most severe worldwide problems for agriculture is little rainfall. About 4/10th of the world's agricultural land lies in arid and semiarid regions, cultivating less water-demanding crops like millets, pulses, and oilseeds. These climate-resilient native rice varieties have long vertical roots, and no lateral ones, with the least leaf curling (drought stress). Plants of the short duration variety normally are drought-tolerant to some extent. Navdanya is conserving 39 drought-tolerant rice varieties in Odisha.

Drought Resistant Aromatic and Therapeutic Rice Varieties

There are two other unique rice varieties, Differently Aromatic rice varieties (plenty) and Therapeutic (medicinal) rice varieties (few). Aromatic rice varieties can sustain in water deficit conditions (semi drought), unlike other paddy varieties. Therapeutic rice varieties also survive drought to a considerable extent. Navdanya has conserved 55 aromatic and two therapeutic rice varieties in Odisha. These varieties have been produced through Darwinian factors, both as natural selection as well as artificial selection with mutation over centuries.

Climate resilient Odisha rice varieties have performed exceedingly well on introduction in disaster areas such as Ersama in Odisha, Nagapattinam in Tamil Nadu, and Indonesia, especially regarding their tillering behaviors: 10 in Balasore, 14 in Ersama, 35 in Nagapattinam, and 54 in Indonesia (the last two under the SRI method of cultivation).

Currently, in Odisha, we are conserving 804 varieties of native paddy and of these, 184 varieties are climate resilient. Hundreds of quintals of seeds of flood and salt-tolerant diverse rice landraces from Navdanya's Odisha Seed bank and seed keepers have been provided to disaster-hit farmers in post-Orissa super cyclone at Ersama and Astarang in Odisha, post-Indian ocean tsunami at Nagapattinam in Tamil Nadu, Nandigram in West Bengal, and also to Indonesian farmers. In 2013, after the massive destruction of the standing rice crop in coastal Odisha by cyclone Phailin, Navdanya also distributed 20 flood and salt-tolerant Indigenous rice seeds to farmers of Balasore and Mayurbhanj districts.

Navdanya was able to save climate resilient seed varieties throughout the country. Over the last two decades of our experience working with farmers in different ecologies, the farmers needed to use native seeds that require much less water and are also resilient to diverse environments and capable of withstanding in different climatic stresses.

Navdanya, in August 2006, established seed banks in Jaisalmer (drought-resistant crops), and Orissa (saline, drought, and flood-resistant rice) to help with various dimensions of preparedness in the face of extreme climate changes, like the floods in Barmer (Rajasthan). In 2007, Navdanya established a seed bank in the village Bajkul, the disaster-hit Nandigram block in Midinapur district of West Bengal. In these community seed banks, Navdanya is saving and multiplying Indigenous climate-resilient varieties of different crops. We are currently multiplying seeds of cereals, millets, pseudocereals, pulses, oilseeds, fruits, and vegetables.

In Odisha, the seasons have become unpredictable; the frequency of rains, droughts, and saline inundations have increased substantially. Consequently, paddy is affected, but is more so with the so-called hybrids and high-yielding varieties. However, the climate-adapted rice varieties have evolved to naturally sustain the impacts of climate change and maintain yield. Research currently being carried out in India and abroad to develop climate tolerant rice varieties is unnecessary. Conservation and propagation of the climate-adapted varieties is necessary.

Some varieties are more able to withstand complete submergence for days. A gene named "sub IA" has been identified in these rice varieties. Such genes have been evolved naturally in these rice varieties which are cultivated in predominantly submerged coastal flood plains of Orissa where the crop plants remain wholly under water for days, yet survive to hand over a good yield.

Orissa is endowed with some drought-tolerant rice varieties, a few of which are of high therapeutic importance. Drought stress crops exhibit inhibition of lateral root development as an adaptive response to the stress. The drought response is mediated by a gene that produces the phytohormone, "abscisic acid" which prevents lateral root development. Drought tolerant rice varieties do not exhibit much tillering and are of shorter day durations.

The rice as a crop was brought from the arid areas to the coastal plains centuries ago. The tall indica rice varieties, thus evolved, have the ability to survive submergence.

Seeds of Hope in Natural Disasters

The Seeds of Hope (*Asha Ke Bija*) Program of Navdanya aims to provide an emergency supply of Indigenous varieties of seeds to those who need them and have lost their local varieties due to natural disasters or the Green Revolution policy of the government.

■ **Orissa super cyclone, 1999**

During the Odisha super cyclone in 1999, Navdanya provided victims with a total of 100 quintals of paddy seeds of 14 varieties of native and nativized paddy to three devastated villages; namely Talang, Dharijan, and Junagari under Gadabishnupur GP in Ersama block of Jagatsingpur district. On May 27, 2000, Navdanya provided additional support through the Chachakhai Yubak Sangh, and one such village, Manduki under Astranga block of Puri district on May 28, 2000. Other than paddy, native vegetable seeds were also given to the farmers and district administration for free distribution.

■ **Tsunami, 2004**

During the 2004 Tsunami, Navdanya Odisha gifted 100 quintals saline-resistant native paddy of three varieties to the Joint Director of Agriculture, Nagapattinam, Tamilnadu for free distribution on July 9, 2005 at Nagapattinam.

■ **Sartha, 2007**

During 2007, we distributed 10 quintals of eight saline-resistant paddy varieties among 80 deluged families of Sartha Panchayat under Sadar Balasore block at the Mangrove Field Office, Sartha.

■ **Phailin, 2013**

After the massive destruction of the standing rice crop in coastal Odisha in 2013 by cyclone Phailin, Navdanya distributed 100 quintals of 20 flood and salt-tolerant Indigenous rice seeds to 400 farmers of Balasore and Mayurbhanj districts.

■ **Nandigram, 2007**

In 2007, Navdanya established a seed bank in the village Bajkul in the disaster hit Nandigram block in Midinapur district of West Bengal with 10 quintals of five saline resistant native paddy varieties through the Taj Group of volunteers led by Sk. Ahmmad Uddin.

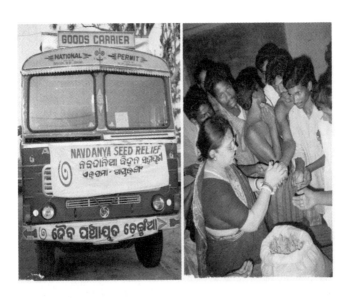

■ **Nepal earthquake, 2015**

On April 25, 2015, an earthquake of 7.6 Richter's struck Nepal. The aftershocks followed, and a second quake measuring 7.3 Richter's struck on May 12, killing over 9000 people. Navdanya provided about 2000 farmers with seeds of paddy, maize, millets, and vegetables.

Climate resilience depends on our saving and spreading the seeds of hope, seeds of freedom, and seeds of resilience.

SECTION 5   Biodiversity for Pest Control: Managing Pests
without Pesticides

According to a UN report, 200,000 people die from acute pesticide poisoning each year, with 99% of cases occurring in developing countries "where health, safety and environmental regulations are weaker and less strictly applied."

The UN's Special Rapporteur Hilal Elver stated that the "[r]eliance on hazardous pesticides is a short-term solution that undermines the rights to adequate food and health for present and future generations," in a recent report on the right to food. Pesticides have been linked to many health conditions, such as Parkinson's disease, Alzheimer's disease, cancers, hormone disruption, fertility problems, respiratory diseases, and more.

Agrarian economies such as Punjab and Kerala are suffering the brunt of indiscriminate use of these synthetic chemicals. Glyphosate and Endosulphan are frequently used synthetic chemicals, which are marketed to farmers as "medicines."

The consequences of these are still evident in the regions in India where they were indiscriminately used. Endosulphan was used in the cashew plantations of Kerala in the Kasargod district for 20 years from the mid-1970s. The consequences of these were seen in the villagers and the wildlife surrounding the areas. Endosulphan affects the endocrine and the genetic systems of people as well as animals. It can also impact reproductive development, sensory losses, neurotoxicity, endocrine disruption, long-term contamination, bioaccumulation, and autism.

Punjab, the first region where the Green Revolution was introduced in India, has been witnessing the consequences of the indiscriminate use of chemicals such as synthetic pesticides, fungicides, and fertilizers. The incidences of cancer increased, which was related to the ascending use of chemicals. There is a daily passenger train that runs from Bhatinda in Punjab to Bikaner in Rajasthan, which has been named the "cancer train." It carries the cancer patients who are victims of the indiscriminate use of chemicals for treatment to a charitable hospital in Bikaner.

The metaphor for pesticide use in agriculture then becomes war.

🌿 VANDANA SHIVA'S LIFE-LONG BIODIVERSITY JOURNEY

My own biodiversity journey in agriculture began with the Bhopal genocide in 1984, when thousands were killed on the night of December 2. Thousands of children still continue to be born maimed.

That is the day I started to look at where pesticides came from and realized they were war chemicals. Bhopal is also the reason I started the Neem campaign to promote nonviolent methods of pest control.

Neem patent (0436257 B1) was granted to the United States Department of Agriculture and the multinational corporation W.R. Grace to control fungi on plants by the aid of an extract of seeds from the neem tree.

The patenting of the fungicidal properties of Neem was a blatant example of biopiracy and indigenous knowledge. I joined Magda Alvoet, President of the European Parliament's Green Party, and Linda Bullard, President of the International Federation of Organic Agriculture. I challenged the patent on the grounds of "lack of novelty and inventive step." We demanded the invalidation of the patent, among others, that the fungicide qualities of the Neem and its use have been known in India for over 2000 years and for users to make insect repellents, soaps, cosmetics, and contraceptives and was finally revoked.

On May 10, 2005, the European Patent Office (EPO) revoked the Neem Patent, 11 years after our challenge to biopiracy. Punjab has emerged as the toxic capital of India with half a century of the chemical-intensive green revolution. The monocultures of rice and wheat are a perfect breeding ground for pests. And the use of toxic pesticides has kept escalating in Punjab; while pests are not a problem in ecologically balanced agriculture, in an unstable agricultural system, they pose a series of challenges to agronomy.

There are two reasons why it's wrong to see all insects as "enemies" that have to be killed with lethal chemical weapons. Firstly, it fails to control pests. Secondly, the toxins harm humans. Pesticides have failed to control pests and have led to the emergence of new pests and resistance in old pests, requiring increased pesticide use. Pesticides create pests by destroying the pest predator balance.

Having destroyed nature's mechanisms for controlling pests through the destruction of diversity, "miracle seeds" became mechanisms for breeding new pests and creating new diseases. The treadmill of breeding new varieties runs incessantly as ecologically vulnerable varieties create new pests, which create the need for breeding yet newer varieties. The only miracle that seems to have been achieved by the Green Revolution is the creation of new pests and diseases and the ever-increasing demand for pesticides. Yet, the costs of new pests and poisonous pesticides were never counted as part of the "miracle" of the new seeds that modern plant breeders have gifted the world in the name of increasing "food security."

Having failed to control pests through the Green Revolution, the pesticide industry has now introduced the second Green Revolution based on genetically engineered seeds, including Punjab where Bt Cotton has been introduced. Bt crops have a gene for producing a toxic introduced with them. The plant itself becomes a pesticide, producing toxins in every cell, all the time. Genetic engineering has also failed as a technology for controlling pests. The bollworm, which it was supposed to control, has evolved resistance, and now pests are emerging every year. The result is a 13-fold increase in pesticide use. Costly seeds and chemicals push farmers into a debt trap, and debt has led to over 300,000 farmers committing suicide in India.

The International Assessment of Agricultural Knowledge, Science and Technology for Development (IAASTD) Synthesis Report is the largest review of our current agricultural systems. This was a multi-stakeholder process that involved over 400 scientific authors and 52 countries. The report concluded that our current food production systems are unsustainable and need to change. It recommended the adoption of ecological systems such as organic agriculture (IAASTD 2008).

Synthetic poisons (pesticides, fungicides, herbicides, and fertilizers) had increased exponentially from when Rachel Carson wrote *Silent Spring* in 1962. The body of science shows that toxic agricultural chemicals are responsible for declines in biodiversity and other environmental health problems. These toxic chemicals now pervade the whole planet, polluting our water, soil, air, and most significantly, the tissues of most living organisms (Aldridge 2003; Buznikov 2001; Cabello 2001; Colborn 1996; Hayes 2002 & 2003; Qiao 2001; Short 1994; Storrs 2004; Tilman *et al.*, 2001). There are three reasons why we should not use these poisons:

1. They are destroying the very basis of our food security. Pesticides are killing pollinators such as bees which, according to the UN, contribute nearly $600 billion to the food economy. As a recent German study shows, 75% of insects have disappeared. Another study from France noted the disappearance of bird species. This is the real threat of extinction we face, not the unscientific "War on Bugs" that falsely purports that we are threatened with extinction because insects are eating all our food.

2. There is an unscientific claim that the poisons used as pesticides are "safe." As Will Allen reports, when independent scientists establish harm, they are attacked, as was the case of Dr. Seralini of France. When the World Health Organisation assesses Glyphosate (Roundup) to be a carcinogen, the WHO is attacked. This is an attack on the environment and public health, on science, and on democracy. Claiming safety by attacking scientific knowledge and regulatory institutions is not science. We have called it a crime against nature and humanity at the Monsanto Tribunal we organized in October 2016 in the Hague.

3. Poisons in agriculture are unscientific because they fail as a pest control technology. It fails to understand the ecology of pests and pesticides. It reduces pest management to the violent use of chemicals.

In de Bach's view:

> "The philosophy of pest control by chemicals has been to achieve the highest kill possible. Such an objective, the highest kill possible, combined with ignorance of or disregard for non-target insects and mites is guaranteed to be the quickest road to upset residences and the development of resistance."

A whitefly epidemic devastated 60% of the Bt cotton crop in Punjab in 2016, and farmers had to use 10-12 sprays, each costing Rs 3200. And in addition, there is the high cost of Bt seeds sold by Monsanto Mahyco. In Maharashtra, Haryana, and Punjab, farmers growing non Bt desi cotton had been impacted by pests like those in Bt cotton fields, but organic farmers in Punjab had no whitefly attack. A scientific approach to what is happening in Punjab would draw the inference that pesticides and Bt are creating pests, while non Bt seeds and organic practices are controlling them.

The second step would be to identify the ecological processes that create pests in Bt crops and in fields using heavy doses of pesticides. The third scientifically enlightened step would be to promote effective and sustainable pest control technologies, such as ecological agriculture based on biodiversity, and to stop pushing failed and costly technologies like Bt and pesticides. Ecological science teaches us that pests are created by industrial agriculture through the following processes:

+ The promotion of monocultures
+ The chemical fertilization of crops, making plants more vulnerable to pests
+ The emergence of resistance in pests by the spraying of pesticides
+ The killing of friendly species which control pests, thus disrupting the pest-predator balance
+ GMO Bt cotton, which is engineered to produce a Bt toxin in every cell of the plant, which makes the plant vulnerable to attack by non-target insects and contributes to the emergence of resistance in the bollworm

Insects are the dominant life form on Earth. Millions may exist in a single acre of land. About one million species have been described, and there may be as many as ten times that many yet to be identified. Of all creatures on Earth, insects are the main consumers of plants. They also play a major role in the breakdown of plant and animal material and constitute a major food source for many other animals.

Insects are extraordinarily adaptable creatures, having evolved to live successfully in most environments on earth, including deserts and the Antarctic. The only place where insects are not commonly found is the oceans. If they are not physically equipped to live in a stressful environment, insects have adopted behaviors to avoid such stresses. Insects possess an amazing diversity in size, form, and behavior.

It is believed that insects are so successful because they have a protective shell or exoskeleton, they are small, and they can fly. Their small size and ability to fly permit escape from enemies and dispersal to new environments. Because they are small, they require only small amounts of food and can exist in very small niches or spaces. In addition, insects can produce large numbers of offspring relatively quickly. Insect populations also possess considerable genetic diversity and a great potential for adaptation to different or changing environments. This makes them an especially formidable pest of crops, able to adapt to new plant varieties as they are developed or rapidly becoming resistant to insecticides.

Insects are directly beneficial to humans by producing honey, silk, wax, and other products. Indirectly, they are important as pollinators of crops, natural enemies of pests, scavengers, and food for other creatures. At the same time, insects are major pests of humans and domesticated animals because they destroy crops and spread vector diseases. In reality, less than one percent of insect species are pests, and only a few hundred of these are consistently a problem. In the context of agriculture, an insect is a pest if its presence or damage results in an economically important loss.

5.1 *Insect Reproduction, Metamorphosis, and Ecology*

Most species of insects have males and females that mate and reproduce sexually. In some cases, males are rare or present only at certain times of the year. In the absence of males, females of some species may still reproduce. This is common, particularly among aphids. In many species of wasps, unfertilized eggs become males, while fertilized eggs become females. In a few species, females produce only females.

A single embryo typically develops within each egg, except in the case of polyembryony, where hundreds of embryos may develop per egg. Insects may reproduce by laying eggs or, in some species, the eggs may hatch within the female, which soon deposits young. In another strategy common to aphids, the eggs hatch within the female, and the immatures remain within the female for some time before birth.

Growth and Development (Metamorphosis)

Insects typically pass through four distinct life stages: egg, larva or nymph, pupa, and adult. Eggs are laid singly or in masses, in or on plant tissue or another insect. The embryo within the egg develops, and eventually, a larva or nymph emerges from the egg. There are generally several larval or nymphal stages (instars), each progressively larger and requiring a molt, or shed of the outer skin, between each stage. Most weight gain (sometimes > 90%) occurs during the last one or two instars. In general, neither eggs, pupae, nor adults grow in size; all growth occurs during the larval or nymphal stages.

Ecology

Ecology is the study of the interrelationships between organisms and their environment. An insect's environment may be described by physical factors such as temperature, wind, humidity, light, and biological factors such as other members of the species, food sources, natural enemies, and competitors (organisms using the same space or food source). An understanding or at least an appreciation of these physical and biological (ecological) factors and how they relate to insect diversity, activity (timing of insect appearance or phenology), and abundance is critical for successful pest management.

Some insect species have a single generation per season (univoltine), while others may have several (multivoltine). The striped cucumber beetle, for example, overwinters as an adult, emerges in the spring, and lays eggs near the roots of young cucurbit plants. The eggs hatch, producing larvae that emerge as adults later in the summer. These adults overwinter to start the cycle again the next year. In contrast, egg parasitoids like *Trichogramma* overwinter as immatures within the egg of their host. During the summer, they may have several generations.

Insects adapt to many types of environmental conditions during their seasonal cycle. To survive the harsh winters, cucumber beetles enter a dormant state. While in this dormant state, metabolic activity is minimal, and no reproduction or growth occurs. Dormancy can also occur at other times of the year when conditions may be stressful for the insect.

It is often better to consider insects as populations rather than individuals, especially within the context of an agroecosystem. Populations have attributes such as density (number per unit area), age distribution (proportion in each life stage), and birth and death rates. Understanding the attributes of a pest population is important for good management. Knowing the age distribution of a pest population may indicate the potential for crop damage. For example, if most of the striped cucumber beetles are immature, direct damage to the above-ground portions of the plant is unlikely. Similarly, if the density of a pest is known and can be related to the potential for damage, an action may be required to protect the crop. Information about death rates due to natural enemies can be very important. Natural enemies do nothing but reduce pest populations and understanding and quantifying their impact is important to effective pest management. This is even more reason to conserve their numbers.

What is a Pest?

There are thousands of organisms in the world, but we do not consider all of them pests. When the population of an organism reaches a level where it can cause considerable damage to the crop, it becomes a pest. They can be either crop pests or storage pests, depending on whether they destroy crops on the field or during storage. Pest damage is a function of the vulnerability of the crop and the pest population, which is determined by the farm's ecology. Organic crops are less pest prone than chemically produced crops. Diverse crops reduce pest population through pest predator balance, while monocultures increase the vulnerability of pests. What are the pests we are so very concerned about in controlling or destroying using all the means at our command? A variety of animal plant and microbial pests cause a wide range of damage to farms, gardens, landscapes, trees, buildings, humans, pets, and livestock.

For a variety of reasons, human beings wish to control or eliminate these pests, largely because of economic losses they cause through the causation of disease, leading to the destruction of standing crops or even during storage. These pests are equally or sometimes more harmful for humans and wildlife, animals, and to the environment and ecology, including buildings. These pests could be insects, mites, weeds, fungi, bacteria, viruses, rodents, etc., even though, as a class, these are a natural part of our environment.

Less than one out of every one thousand insects are pests. Unfortunately, insecticides kill both pests and natural enemies of these pests indiscriminately, these are referred to as predators. It has been postulated that there are million of species of insects on this Earth, of which some 5,000–15,000 have turned into pests. Many of the other insects, which have the potential to become pests, are kept in check by climate, food, or natural enemies—the predator and the parasite. The introduction of chemical pesticides in the complex interplay of predators, crops, farm animals, and man has raised three key problems:

1. Creating susceptibility to pests by destabilizing the plant metabolism
2. The killing of natural enemies of the target pest leading to an explosion in the population of the original pests
3. Destruction by pesticides of the large number of non-target species of natural enemies, leading to a class of evolution of new secondary pests in the form of resistant varieties of pests

In a biological system, every organism has a niche and is a part of the delicate web of the food system. The spraying of chemicals leads to the mass destruction of beneficial insects such as soil nematodes and pollinating insects, which in turn leads to a reduction in cross-pollination that reduces the genetic base of the region. This, in turn, affects the resilience and ecological amplitude of the ecosystem.

From time immemorial, farmers had the wisdom and knowledge of biological pest management that had been part and parcel of farming in India. The farmers understood the delicate web of nature and understood the intricacies of the food web. Traditional organic farming has normal procedures of growing diverse crops, which is the basis of ecological pest management.

5.2 *Recognizing the Role of Natural Enemies*

In nature, we find that every pest has a predator (an organism that feeds on the pest), which helps to keep the pest population in check. A sudden decrease in the predator population could lead to an increase in the pest population, causing extensive damage to crops. The predator and prey populations are so interdependent that an increase or decrease in either population causes drastic changes in the population of the other.

Natural enemies play an essential role in limiting the densities of potential pests. This has been demonstrated repeatedly when pesticides have devastated the natural enemies of potential pests. The non-toxic methods to control a key pest, the reduced use of pesticides, and the increased survival of natural enemies frequently lessen the numbers and damage of formerly important secondary pest species.

Applying a chemical insecticide has several direct and indirect effects: primarily, it kills pests, thereby immediately reducing their population size. But there are then indirect effects that increase pest abundance.

Pesticides kill the predators of pests, thereby indirectly benefiting the pests. The pests rebound, and they rebound to a higher density than previously expected because their equilibrium density is increased. *Ex. Bollworm in cotton.*

- It kills the predators of other herbivorous insects that were not yet pests, thereby allowing these insects to reach higher densities and become pests. The following table depicts the data showing pest incidence in cotton.

Table 1: Increase in pests incidence due to pesticide applications in cotton in Nicaragua

YEAR	NUMBER OF PEST SPECIES	NUMBER OF PESTICIDE APPLICATIONS	RELATIVE CROP YIELD
1950	2	0–5	100%
1955	5	8–10	80%
1965	8	25–30	70%
1979	24	50–60	–

5.3 *Pesticide Resistance*

Pesticide selects for pesticide resistance in the pest populations. Herbivorous insects already have evolved ways of overcoming toxins produced by plants and are able to evolve means of detoxifying or avoiding pesticides quickly.

These effects combine to put the farmer on an escalating pattern of applying more and more pesticides and more kinds of pesticides to control more and more pests. It is called a "pesticide treadmill," but it really isn't a treadmill because the farmer is continually losing ground. Pesticide application is really like an addiction to narcotics, in that once started creates its own demand.

The three categories of natural enemies of insect pests are:

1. Predators
2. Parasitoids
3. Pathogens

Predators

Many different kinds of predators feed on insects. Insects are an important part of the diet of many vertebrates, including birds, amphibians, reptiles, fish, and mammals. These insectivorous vertebrates usually feed on many insect species and rarely focus on pests unless they are very abundant. Insect and other arthropod predators are more often used in biological control because they feed on a smaller range of prey species and because arthropod predators, with their shorter life cycles, may fluctuate in population density in response to changes in the density of their prey. Important insect predators include lady beetles, ground beetles, rove beetles, flower bugs, true predatory bugs, lacewings, and hoverflies. Spiders and some families of mites are also predators of insects, pest species of mites, and other arthropods.

Parasitoids

Parasitoids are insects with an immature stage that develops on or in a single insect host and ultimately kill the host. The adults are typically free-living, and maybe honeydew, plant nectar, or pollen. Because parasitoids must be adapted to their hosts' life cycle, physiology, and defenses, they are limited in their host range, and many are highly specialized. Thus, accurate identification of the host and parasitoid species is critically important in using parasitoids for biological control.

Pathogens

Bacteria, fungi, protozoans, and viruses that cause disease infect insects and plants. These diseases may reduce the rate of feeding and growth of insect pests, slow or prevent their reproduction, or kill them. In addition, insects are also attacked by some species of nematodes that, with their bacterial symbionts, cause disease or death. Under certain environmental conditions, diseases can multiply and spread naturally through an insect population, particularly when the density of the insects is high.

5.4 *Predators Occurring in the Field*

Lady beetles

These familiar creatures, in both larval and adult stages, feed on soft-bodied insects, especially aphids. You can attract them by planting nectar plants (nectar is an alternate food source) and those that attract aphids. These include alyssum, legumes, and flowers in the Umbelliferae family (dill, wild carrot, fennel, yarrow, and so on).

Target pests

Aphids, leafhoppers, scales, mites, mealybugs

Parasitic or predatory wasps

Encarsia formosa are small wasps that parasitize greenhouse whiteflies.

Trichogramma wasps parasitize eggs of leaf-eating caterpillars such as cabbage loopers.

Target pests

Caterpillars, aphids, mealybugs, leafhoppers, greenhouse whiteflies.

Praying mantis

They can be wonderful allies for gardeners (and great fun to watch), but they eat such a variety of insects that you wouldn't want to use them for an outbreak of any one pest.

Target Pests

Most pest insects and eggs.

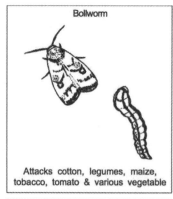

Bollworm

Attacks cotton, legumes, maize, tobacco, tomato & various vegetable

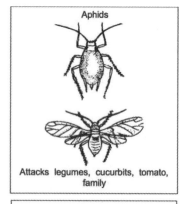

Aphids

Attacks legumes, cucurbits, tomato, family

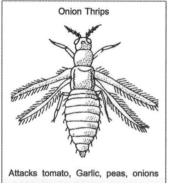

Onion Thrips

Attacks tomato, Garlic, peas, onions

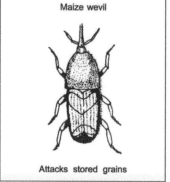

Maize wevil

Attacks stored grains

Wasp parasiticing on aphid

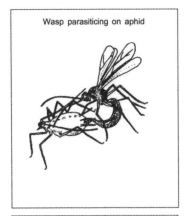

Soldier beetle

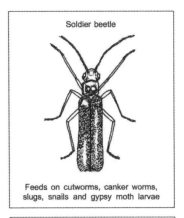

Feeds on cutworms, canker worms, slugs, snails and gypsy moth larvae

Praying mantis

Eats all small insects

Lady bird beetle

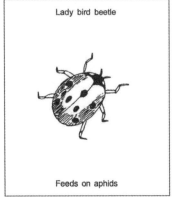

Feeds on aphids

Syrphid fly

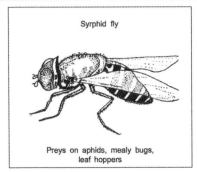

Preys on aphids, mealy bugs, leaf hoppers

5.5 Non-Chemical Methods of Pest Control

The range of non-chemical options available may vary with the pest species, pest intensity or severity, and effectiveness of the option. Several key non-chemical options that may help reduce the amount of pesticides used in and around homes are listed below.

Exclusion: Any measure used to prevent entry of organisms in the farm field by digging trenches.

Sanitation: Maintaining clean surroundings in the farm where pests can feed, breed, and hide. Sanitary measures include cleanliness in and around the farm by discarding plastics, or any other inorganic substance. The on-farm refuge should be transferred to the compost bins that are constructed near the farm fields.

Habitat modification: Creating a live barrier around the perimeter of the farm fields that will reduce the incidence of many ground-dwelling pests as the allelopathic effects of the roots will ensure that the pathogens are not freely invading the farm soil. A suitable example is the *Prosopis juliflora* live fencing in the bunds of agriculture fields in Rajasthan.

Mechanical control

A bin with tweezers is the best mechanical tools used for killing visible and less mobile or immobile pests. On infested plants, hand-picking insects (e.g., hornworms) is a partially effective means of pest control. Infested leaves must be excised from plants, bagged, and discarded.

Traps are escape-proof devices that capture highly mobile and active pests. Colored (yellow) sticky traps are effective in capturing whiteflies and aphids. Sticky traps can be baited with commercial lures (pheromones and food attractants) to enhance trap catch. These methods are used in places where the pests have a crawling feature of movement.

Traps are useful for early detection and continuous monitoring of infestations. They are not effective in reducing populations unless the pest population is isolated or confined to a small area. The chance of detecting the presence of pests in a given area is related to the number of traps used. Therefore, when pests are present in very low numbers, it is advantageous to use more than a few traps. Pests must be active or mobile to be captured in traps. Therefore, any environmental variable (temperature, humidity, wind, light, or food) or biological factor (age, sex, mating status, etc.) that influences pest activity affects trap catch.

Biological Control Agents

Parasitic and predatory insects, mites, and nematodes are now commercially available to control pests. For example, lacewing larvae and ladybird beetle larvae, and adults are predators of aphids. Parasitic and predatory organisms should be used only where pesticides are discontinued or were not previously used because these beneficial organisms are highly susceptible to pesticides.

Natural environments tend to be balanced environments, where organisms depend on one another and also constrain one another by competition for resources or by parasitism, predation, etc. But human influences can upset these balances, and this is most evident when an exotic organism is introduced on purpose or by accident. Many of the most serious pests, crop diseases, or invasive weeds are the result of "introductions" from foreign lands. The newly introduced organisms find a favorable environment, free from their previous constraints, and they proliferate to achieve "pest" status. Entomologists have a useful term for this—they refer to the constraining elements in the region of origin as "the natural enemy complex."

We can define Biological Control (biocontrol) as the practice or process by which an undesirable organism is controlled by means of another (beneficial) organism.

In other words, biocontrol is both a naturally occurring process (which we can exploit) and the purposeful use of one organism to control another. In practice, biocontrol can be achieved by three methods.

+ Inundative release (also termed "classical biocontrol") in which a natural enemy of a target pest, pathogen, or weed is introduced to a region from which it is absent to give long-term control of the problem.

+ Biopesticide approach in which a biocontrol agent is applied when required (often repeatedly) in the same way as a chemical control agent is used. Examples of this include the use of *Bacillus thuringensis*, *Phoebiosis gigantean*, and *Agrobacterium radiobacter*.

+ Management and manipulation of the environment to favor the activities of naturally occurring control agents. Biological control is the use of living organisms to suppress pest populations, making them less damaging than they would otherwise be. Biological control can be used against all types of pests, including vertebrates, plant pathogens, weeds, and insects, but the methods and agents used are different for each type of pest.

Three Primary Methods of Using Biological Control

+ Conservation of existing natural enemies
+ Introducing new natural enemies and establishing a permanent population (called "classical biological control")
+ Mass rearing and periodic release, either on a seasonal basis or inundatively.

Biological control aims to reduce or even eliminate the use of chemicals to control pests, diseases, and weeds related to plants of economic interest and develop and apply specific products that are not dependent on foreign technologies and imported raw materials.

The biological methods of pest control are economically sound, environmentally safe, preserve the health of rural workers and their families, and provide healthier food.

One of the mainstays of integrated pest management is the use of crop varieties that are resistant or tolerant to insect pests and diseases. A resistant variety may be less preferred by the insect pest or adversely affect its normal development and survival, or the plant may tolerate the damage without an economic loss in yield or quality. Disease-resistant vegetables are widely used, whereas insect-resistant varieties are less common but important. The best method is to use the Indigenous seeds of the locality, which results from years of selection of the farmer.

Advantages of this tactic include ease of use, compatibility with other integrated pest management tactics, low cost, and cumulative impact on the pest (each subsequent generation of the pest is further reduced) with minimal environmental impact. However, the development of resistant or pest-tolerant plant varieties may require considerable time and money, and resistance is not necessarily permanent. Just as insect populations have developed resistance to insecticides, populations of insects have developed that are now able to damage plant varieties that were previously resistant.

Cultural Control

There are many agricultural practices that make the environment less favorable to insect pests. Examples include cultivation of alternate hosts (e.g., weeds), crop rotation, selection of planting sites, trap crops, and adjusting the timing of planting or harvest.

Crop rotation, for example, is highly recommended for the management of the Colorado potato beetle. Over winter in or near potato fields, the beetles require potato or related plants for food when they emerge in the spring. With cool temperatures and no suitable food, the beetles will only crawl and be unable to fly. Planting potatoes well away from the previous year's crop prevents access to needed food, and the beetles starve. The severity and incidence of many plant diseases can also be minimized by crop rotation, and selection of the planting site may affect the severity of insect infestations.

Trap crops are planted to attract and hold pest insects where they can be managed more efficiently and prevent or reduce their movement onto valuable crops. Early planted potatoes can act as a trap crop for Colorado potato beetles emerging in the spring. Since the early potatoes are the only food source available, the beetles will congregate on these plants where they can more easily be controlled. Adjusting the timing of planting or harvest is another cultural control technique.

5.6 *The Importance of Integrated Pest Management (IPM)*

Integrated Pest Management is a method of pest control that keeps the environment safe and is ecologically sound and economically viable.

The main strategies of the IPM are:
- Tilling the soil
- Timing of sowing, planting, and harvesting
- Crop rotations
- Destroying crop residues of infested plants
- Using resistant varieties
- Using good quality seeds

In other words, IPM combines biological and agronomic approaches, making up a strategy that is not only sustainable over a time frame but also least damaging to the environment.

5.7 *Indigenous Methods of Vrikshayurveda (Pest Control)*

India has a rich tradition of rigorous study of plant diseases. The compendium of plant medicine is called *Vrkshayurveda*. This compendium records diverse methods of ancient treatments against insect attacks. Some of the treatments are as follows:

- Nutmeg (*Asafoetida*) is mixed with two kinds of sweet flag (*Vacha*), pepper (*Erycibe paniculata*), marking nut (*lappiga aliona*), mustard and paste of cow's horn. It is mixed in cow's urine and applied around the trees or plants.
- Fumigation of cow's horn, marking nut, neem, nutgrass (*Musta*), sweet flag (*Vacha*), viranga (*Vidanga*), Aconite (*Atibhisa*) and Indian beech (*Karanja*) in conjuction with resin of sal tree (*Aarjarsa*), white mustard (*Siddhartha*) and five-leaved Chaste tree (*Sinduvara*) destroys insects of trees.
- Use of tobacco as a natural pesticide.
- Use of buttermilk in controlling diseases of cotton. Buttermilk is kashaaya (astringent) and amla (sour) and is also a digestive.

5.8 *Traditional Techniques for Prevention of Pest Attack*

Grain/seeds may be periodically dried in the sun. This chases away adult insects. However, eggs and larvae may still remain.

Storage rooms may be smoked regularly with neem leaves to keep away moths, weevils, and beetles.

Wood, cow dung ash, and sand may be mixed with grain. One effect of adding these is that they fill the inter-granular spaces and, therefore, restrict insect movement.

Adding inert mineral dust and special types of clay (including activated charcoal and heat-activated clay dust) to the grain is also practiced. These scratch the thin waterproofing layer, which exists on the outside surface of the insect's body wall, causing a loss of water and its death from desiccation. Wood ash and sand can also have this effect.

A small clay lamp filled with oil may be lit and placed inside the storage container before it is sealed. The lamp will burn until the oxygen in the container is exhausted; this will lead to the insects also dying from lack of oxygen.

Dirt or cow urine is often sprayed to keep away insects. Some inorganic and organic pesticides are traditionally used for the control of major crop pests and pathogens.

Recipe for Vegetable and Pulse Pests

Common pests and pathogens of pulses or vegetables include bacterial blight, aphids, fungal and leaf spots, as well as viral disease of chili, insect attack on pulses, lemon, and watermelon. Others include, heliothis and Hairy caterpillars and aphid attack on *Foeniculum vulgar*.

Ingredients and Materials

- Leaf extracts of neem and tobacco (5:1)
- Cattle urine (2-3 days old)
- Calcium oxide and wood dust mixture
- Leaf extracts of Vilayati Babul (*Prosopis juliflora*) mixed with water and (3lt. extract/acre)
- Aquatic solution of alum
- Extract of *Bassia latifolia* or *Pongamia glabra*
- 1 kg of garlic, crushed and soaked overnight in 200 ml of kerosene
- 2 kg of ground green chili and 200 lt of water for spraying
- Dry leaves of eucalyptus species, green as well as dry are burnt in the early morning when wind velocity is not high

Recipe for Storage Pests

Storage pests including paddy, pulse, groundnut, leaf rollers and hoppers, whiteflies, leaf roller, and gundhi bug, aphids, stem borer and bacterial diseases, and termites in sugarcane and coconut plant.

Ingredients and Materials

- Leaf powder of *Vitex negundo*
- Split seed coat pieces of cashew (*Anacardium occidentale*), leaves of *Moringa pterygosperma*.
- Camphor leaves of *Sphaeranthus indicus*. Leaf extract of *Lasiosiphon eriocephalus*. Mixture of Asaphoetida and cattle urine or mixture of garlic, chilli and nutmeg in water.
- Neem cake

- Dry cow dung and vegetable waste brunt before planting of the crop, in furrows opened for planting sugarcane cuttings.
- Coal tar applying to the lower part of the stem.
- To increase soil fertility, we can take 10 kilos of cow dung and add 250 gm of ghee, stir for 4 hrs. Add 500 gm of honey of 1 kg of jaggery, stir for 4 hours. After, it becomes very good food for soil microorganisms. To this mixture, add 200 liters of water. It is known as *Amrit pani/Sanjivani pani*. Apply it to one acre of land. Then mulch it. Fourteen hundred farmers of Maharashtra, Goa are using this method to increase their wealth of earthworms in the soil. A farmer Pandharpur in Maharashtra has a 23-acre vineyard, where he is using this method. His farm yielded grapes to a tune of one tons per acre, which is a record.
- *Kunwarji Bhai Zadav, All India Kisan Sabha*

5.9 *Biopesticides*

Biopesticides are certain types of pesticides derived from natural materials like animals, plants, bacteria, and certain minerals. They are usually inherently less toxic than conventional pesticides and generally affect only the target pest and closely related organisms, in contrast to broad-spectrum conventional pesticides that affect organisms as different as birds, insects, and mammals. They are also often effective in very small quantities and often decompose quickly, thereby resulting in lower exposures and largely avoiding the pollution problems of conventional pesticides. Biopesticides are safer for humans and the environment than conventional pesticides. They present no residue problems because they disintegrate in nature very rapidly.

Plants as Biopesticides

▧ CORIANDER (*Coriandrum sativum*)

Coriander is an annual herb that grows up to 1–3 feet in height. It is generally grown as a rainfed crop either in pure stands or mixed with other crops. In certain areas, it is grown as an irrigated crop. The leaves, seeds, and oil are used for pest control.

Pest controlled: Aphids

■ GINGER *(Zingiber officinalis)*

Ginger is a perennial herb reaching up to 90 cm in height. Rhizomes are thick, lobed, and yellow in color. Ginger requires a warm and humid climate. It is propagated by seed rhizomes, which are used for pest control.

Pests controlled: American bollworm, aphids, mango anthracnose, pulse beetles, root knot nematode, whitefly, yellow vein mosaic, etc.

■ LEMON GRASS *(Cymbopogan citrates)*

Lemongrass is a perennial grass that grows in tufts. It grows well in mountainous areas. It acts as a repellant and growth disrupter. The roots, leaves, seeds, and oil are used for pest control.

Pests controlled: Fruit flies, mites, mosquitoes, and storage pests.

■ NEEM *(Azadirachta indica)*

Neem is an evergreen tree growing up to an average of 18 m in height. Leaves, seeds, cake, and oil extracts could be prepared for spraying. It acts as a feeding, deterrent, oviposition deterrent, and insect growth regulator.

Pests controlled: Aphids, brown plant hopper, diamond black moth, green leafhopper, root knot nematodes, termites, stem borers, etc.

■ ONION *(Allium cepa)*

Onion is a bulbous biennial and can be cultivated throughout India. Onion bulbs are used as extracts for pest control.

Pests controlled: Nematodes pulse beetle, ticks, tobacco mosaic virus, etc.

■ TOBACCO *(Nicotiana tabacum)*

Tobacco is stout annual with a thick erect stem and few branches and propagates through seeds. The leaves, stalk, and stem can be used for pest control.

Pests controlled: Aphids, Citrus leaf miner, Rice stem borer, Mites, etc.

■ TURMERIC *(Curcuma longa)*

Turmeric is a perennial herb with a short stem and tufted leaves and is propagated by rhizomes. The rhizome extract can be used for pest control.

Pests controlled: Armyworm, aphids.

■ **Garlic** (*Allium sativum*)

Garlic is a hardy perennial, attaining a 30-100 cm. The bulbs, leaves, flowers, and oil are used for pest control.

Pests controlled: Aphids, armyworms, bacteria, colorado beetle, mites, root knot nematode, rice blast fungi, etc.

Methodologies to Prepare Biopesticides

Neem has attracted worldwide attention in recent decades mainly due to its bioactive ingredients that find increasing use in modern crop and grain protection. Research has shown that neem extracts have an effect on nearly 200 species of insects. It is significant that some of these pests are resistant to pesticides or are inherently difficult to control with conventional pesticides (floral thrips, diamondback moth, and several leaf miners). Most neem products belong to the category of medium to broad-spectrum pesticides, i.e., they are effective over a wide range of pests.

A range of neem products such as neem leaf extract, neem seed kernel extract, neem cake extracts, neem oil emulsion, and also neem in combination with other plant extracts for the control of a variety of pests. The technologies using neem are simple, and the farmer can make these products in their own backyard. They have been tested in the farmers' fields and are proven to be effective in controlling a wide range of pests. They have also been used in controlling stored grain pests. To control rats, pieces of papaya fruit are spread near the bunds of the field. Papaya has a chemical substance which causes tissue damage in the mouth of the rats feeding on it.

Preparation of Neem Kernel Extract

+ 50 grams of neem kernel are required for use in 1 liter of water.
+ Pounded gently in such a way that no oil comes out. The outer coat is removed before pounding. This is used as compost.
+ Put the pounded seeds into a muslin cloth and soak overnight in a liter of water.
+ Squeeze the pouch, and the extract is filtered.
+ Add 1ml non-detergent-based soap (*khadi* soap) solution to the filtrate. This acts as an emulsifier.
+ 10 milliliters of emulsifier is added to 1 liter of water. The emulsifier helps the extract to stick well to the leaf surface.

The kernel extract should be milky white and not brownish. The kernel extract does not control sucking insects like aphids, whiteflies, and stem borers. In these cases, one could use the neem oil spray solution.

■ NEEM LEAF EXTRACT

For 5 liters of water, 1 kg of green neem leaf is required. Since the quantity of leaves required for the preparation of this extract is quite high (nearly 80 kg is required for 1 hectare), this can be used for nursery and kitchen gardens. The leaves are soaked overnight in water. The next day, they are ground, and the extract is filtered. The extract is suited for use against leaf-eating caterpillars, grubs, locusts, and grasshoppers. To the extract, the emulsifier is also added.

The advantage of using neem leaf extract is that it is available throughout the year. There is no need to boil the extract since boiling reduces the Azadirachtin content. Hence the cold extract is more effective. Some farmers prefer to soak the leaves for about one week, but this creates a foul smell.

■ NEEM CAKE EXTRACT

A hundred grams of neem cake are required for 1 liter of water. The neem cake is put in a muslin pouch and soaked in water overnight. It is then filtered, and an emulsifier is added at the rate of 10 milliliters for 1 liter of water, after which it is ready for spraying.

■ NEEM OIL SPRAY

Thirty milliliters of neem oil is added to the emulsifier and stirred well to ensure that the oil and water can mix well. After this, 1 liter of water is added and stirred well. It is essential to add the emulsifier with the oil before adding water. It should be used immediately; otherwise, oil droplets will start floating. A knapsack sprayer is better for neem oil spraying than a hand sprayer. (Ecological Management of Pest 129)

■ PONGAM, ALOE, AND NEEM EXTRACT

One kilogram of pounded pongam cake, 1 kg of pounded neem cake, and 250 g of pounded poison nut tree seeds are put in a muslin pouch and soaked overnight in water. In the morning, the pouch is squeezed, and the extract is taken out. This is mixed with 1/2 liter of aloe vera leaf juice. To this, 15 liters of water are added. This is again mixed with 2-3 liters of cow's urine.

Before spraying, 1 liter of this mixture is diluted with 10 liters of water. For an acre, 60-100 liters of spray are used. This is effective in the control of pests of cotton and crossandra.

■ Custard Apple, Neem, Chili Extract

Five hundred milliliters of water are added to 2 kg of ground custard apple leaves and stirred. This is filtered to get the extract, and the filtrate is kept aside. Separately, 500 g of dry fruits of chili is soaked in water overnight. The next day, this is ground and the solution filtered to get the extract. One kilogram of crushed neem fruits is soaked in 2 liters of water overnight, and the extract is filtered. All the three filtrates are subsequently mixed with 50-60 liters of water, filtered again, and sprayed over the crops. *Note:* For the above extracts, 250 milliliters of *khadi* soap solution should be added as an emulsifier before spraying.

■ Pongam or karanj extracts: Leaf extract

+ Soak 1 kg of Pongam leaves in 5 liters of water overnight
+ Grind leaves next morning and filter
+ Add 10 ml of emulsifier (*khadi* soap solution) for every liter of water)
+ Use as spray against leaf-eating caterpillars

■ Pongam or Karanj Sharifa or Sitaphal

+ To spray an acre, 20 kgs of leaves, 100 liters of water, and 100 ml of emulsifier is required

■ Kernel extract

Remove outer coats of seed and pound gently. 50 grams of this is used for 1 liter of water.

+ Place kernel powder in a muslin pouch and soak overnight
+ Squeeze the pouch and filter the extract
+ Add 10 ml of emulsifier for every liter of water and use as a spray
+ To spray an acre, 5 kgs of cake, and 100 liters of water, and 100 ml of emulsifier is required

■ CAKE EXTRACT

+ Take 100 grams of Pongam cake and powder it well
+ Fill a muslin pouch with the powder and soak it overnight in 1 liter of water
+ Squeeze out the pouch and filter the solution
+ Add emulsifier at the rate of 1 ml for every liter. Mix well and use as a spray
+ To spray an acre, 10 kg of Pongam cake, 100 liters of water, and 100 ml of emulsifier is required

■ OIL SPRAY

+ To make 1 liter of spray, 30 ml of Pongam oil is used
+ This was added to the emulsifier (*khadi* soap solution at the rate of 10 ml for every liter of water) and mixed well
+ This solution is added to water and is ready for spraying
+ It is important that the spray be used immediately after it is made
+ To spray an acre, 3 liters of oil, 100 liters of water, and 100 ml of emulsifier is required

■ GARLIC EXTRACT (*Allium sativum*)

+ Use 100 grams finely ground garlic
+ Soak finely ground garlic in 2-tablespoon liquid paraffin for 48 hours
+ Add 30 grams *khadi* soap to 1D 2 liter water and mix well
+ Filter the solution and store it in a plastic container
+ To prepare 1 liter of spray, add 15 ml of extract and mix well
+ To spray an acre, 15 liters of extract and 100 liters of water are required

■ TOBACCO (*Nicotiana tobacum*)

+ Take 250 grams tobacco and boil it in 4 liters of water for 30 minutes
+ Add 30 grams *khadi* soap and mix well
+ Dilute 1 part extract with 4 parts water and use as a spray
+ Adding a little slaked lime increases the potency of the extract
+ This extract is extremely poisonous. Even in very minute quantities it causes death in animals and humans. Do not use sprayed plants for at least 4-5 days after spraying.

5.10 *Seed Treatments*

▪ Panchgan

Material: Cowdung, cow urine, freshmilk, curd, ghee
Quantity: 1 part, 1 part, 1 part, 1/2 part, 1/10 part
For 10 kg of seed: 1 cup, 1 cup, 1 cup, 1/2 cup, 1 spoon
Seedling stage: To dip the seedling; dilute the above proportion in water 5-6 times. (10 times dilution for sugarcane)

+ Use fresh milk with 10% dilution of chilli and tomato to control virus.
+ Use 2% lime water for cut warms: 2% of lime water – keep it for overnight- take out (siphon) clean water-add 2kg ash (white ash)

Seedbed treatment in cabbage:

+ Use 10% fresh milk spray
+ Use ash and turmeric powders; dust it on seedbed at 10 day intervals

Bud sprout in grape/udar beetle:

+ Take 2-3 lit. of fresh milk
+ Add 8 lit. of kerosene
+ Mix 100 lit. of water (per 1 acre)
+ Stir it
+ Spray after 4 pm

▪ Repellent to termites

Use tea powder and agave's leaf extract spray.
Concoction to replace bavistine/carbandizime
Material: Cowdung, cow urine, freshmilk, curd, ghee
Quantity: 40 parts; 40 parts; 6 parts; 5 parts; 1 part

+ Add 0.2% (200gms) yeast
+ Add 0.1% salt
+ Mix it and keep it for 8 days
+ Filter through cloth
+ Dilute it to 10 times. Spray it.

Powdery mildew in grapes

Plant Tulsi and marigold in grape garden, and lemongrass on boarders.

Wet rot in ginger

Keep *Calotropis* (Aak) twigs in irrigation channel.

5.11 *Bird Attractant*

One kilogram of rice and 50 grams of turmeric powder is required to treat an acre. The rice is cooked, and excess water is filtered. This is mixed with turmeric powder. Small lumps of yellow-colored rice are taken in small vessels and placed in the main field in 8 to 10 places. This is kept during the early morning and afternoon. When the birds feed on the rice, they feed on the semi looper larvae prevent in the field. This procedure is repeated until the crop attains the flowering stage, thereby reducing pest attack. This controls pests occurring in rice, namely, the Rice stem borer and armyworm.

1. Spraying should be undertaken in the morning or late in the evening. Under hot conditions, the frequency of spraying should be increased. In winter, spraying every 10 days and every day in the rainy season is recommended.
2. Insects lay eggs on the underside of the leaves. Hence it is important to spray under the leaves also.
3. While using a power sprayer, the quantity of water used should be halved.
4. It is better to use low concentrations of extracts frequently.
5. As a general guideline, it can be said that each acre of land to be protected can be sprayed with 60 liters of ready-to-use solution (not the concentrate). Of course, the volume may have to be varied depending on the exact conditions prevailing, such as the intensity of the pest attack.

The use of insecticidal plants or plant substances in storage protection:
+ The insecticidal plants are helpful in storage protection of many crops.
+ Leaves or neem seeds are stored together with cereals or beans, thus diminishing storage losses.
+ Use powdered rhizomes of the sweet flag (*Acorus calamus*) at a ratio of 1 kg to 50 kg of grain. Mixed well with the grain and applied before storage, it can effectively reduce infestation by important storage pests such as: the rice weevil, khapra beetle, lesser grain borer, and adzuki bean beetle.
+ To protect beans in storage from infestation with bruchids, each kilogram

of beans should be mixed with 2-3 ml of neem oil. It is important to ensure that the oil is well mixed so that each bean is coated. Thus, beans can be protected for six months.

5.12 *Treatment of Stored Grains*

Grains and pulses can be stored by mixing them with neem products like dried leaf powder, kernel powder, or oil. The neem oil used against stored grain pests should be 1% by weight of the grain. If the grain is used for seed purposes, 2% can be used. Using oil is easier than using leaves. The active ingredients of the neem plant are located in their maximum amounts in the seed and kernel.

Storage Pest Control
■ FOR PULSES ALONE

1 kg of any pulse should be coated with 2-5 ml of castor oil before storage. This gets rid of storage pests for 6 months to 1 year.

For vegetables and fruits: Seeds should be treated with wood ash at the rate of 5-10 kg per quintal (100 kg) of seeds before storage.

For cereals: Datura leaf dust should be mixed with grain at a rate of 10 gm per kg of grain.

Wood ash in storage pest control

Wood ash is a very effective pesticide. It is harmless to health. It can be mixed in equal quantity to the total amount of grains. It offers good protection against beetles and other storage pests. Ashes from the leaves of *Lantana* are very effective against pests attacking the sprouts of stored potatoes.

Traditional methods of storage

Neem leaves and Pongam leaves can be spread on the floor where stored grains are kept in gunny bags. In Tanjore, farmers spread these leaves in between the bags at regular intervals.

Add neem leaf powder to the clay soil and swab it on the inner surface of the storage bins. For 1 kg of clay soil, add 10 g of neem leaf powder. After swabbing, store the grains by putting one layer of neem leaves and one layer of grains alternatively. In this way, the stored grains are protected from the pest attack for a year.

Treatment of jute bags for storing grains

The jute bag is dipped into a 10% neem kernel solution (here, no emulsifier need be added to the solution) for 15 minutes. After having been dried in the shade, the bag can now be used for storing grains. The stored grain pests will be repelled by the action of neem.

If the jute bags are new, they should be soaked for half an hour. For jute bags with close meshes and small pores, a thinner solution can be used.

How Neem Works

Neem products such as oil and cake contain a substance called Azadirachtin. The substance reduces the egg-laying capacity of insects. The same substance goes into the plant, and sucking insects cannot feed on the plants. Azadirachtin alters the physiology of insects and breaks their life cycle, reducing the spread of pests.

If seeds or grains are kept inside a house or in a warehouse, where the temperature is stable and sunlight minimal, longer residual action of the neem product is obtained, and the repellent effect will persist for four months.

In storerooms, along with the cow dung that is used for cleaning the mud floor, neem cake or neem oil can be used straight away (in the same concentration as used for spraying purposes). The same could also be used for the mud walls. Neem cake solution or neem kernel extract could also be sprayed. If one is using bamboo bins for storage, then one can paint the bins with a solution prepared from neem cake. To the dry neem cake powder, water is added, and a thick paste of this is painted all over the grain bin. If one wishes to store it for more than four months, the process should be repeated every four months. Neem products work by intervening at several stages of the life cycle of an insect. They may not kill the pest instantaneously, but incapacitate it in a number of ways.

5.13 Crop Disease Management

Disease control measures such as the use of disease-free seed, good crop rotations, and other cultural methods become very important in organic farming situations. For example, increasing the length of crop rotations by including perennial forages for several years can significantly decrease the amount of common root rot inoculum present in the soil. Burying crop residue is not recommended due to the potential for soil erosion and degradation. Crops which are susceptible to similar diseases should not be grown within the recommended number of years of each other for each particular disease.

Weed control is also a key factor in disease management. If weeds that carry disease or are susceptible to it are allowed to persist, crop rotation will not effectively control the disease. For example, suppose the rotation includes a canola crop every fourth or fifth year to avoid sclerotinia stem rot. In that case, susceptible weeds like wild mustard cannot be allowed to proliferate, and susceptible crops such as field peas, field beans, or lentils should not be grown.

5.14 *Protecting Seeds From Insects*

Seeds are commonly spoiled by particular insects, which feed on stored seeds and grains. There are a number of ways to protect seeds in storage.

For every kilogram of seeds to be stored, use half a kilogram of fresh, dry wood ash; a little more ash can be added to cover the seed in the container. Fresh ash, not old ash, should be used, as old ash is usually wet and contaminated with microorganisms. Do not use hot ash, since the seeds may be killed.

Dry, clean sand mixed with the seeds will also provide protection against weevils, as the coarse sand grits will make movement uncomfortable for them.

5.15 *Diversity of Insects and Arthropods at the Navdanya Farm*

Numerous arthropods such as insects and arachnids act as predators of numerous plant disease-causing organisms we usually term as pests. These pests were always present in nature and formed food for the beneficial organisms. Getting rid of them completely may be harmful to the diversity of the beneficial organisms to plants.

To restore balance in the biological functions of birds, insects, and predatory organisms in agroecosystems, we need to retrace our path to manage their populations naturally using an ecosystem approach. Keep a watch on the emergence of diverse insect groups on crops, noting their predation by other life forms such as insects, birds, and reptiles.

Encouraging these natural predators may stand as prominent pest control management in the future. Natural selection and evolution are the mechanisms of nature, and they can go a long way to manage the friendly insects in farmlands. On the contrary, using synthetic chemicals will only filter resistant insects that require continual increase in lethal doses of harmful chemicals. As the wise quoted, "an ounce of prevention is worth a pound of cure" it is worth conserving biological diversity rather than reducing it with synthetic inputs.

Table 2: Beneficial insects and arthropods

FAMILY	SPECIES
Araneidae	*Eriophora sp.*
Oxyopidae	*Oxyopes sp.*
Theridiidae	*Parasteatoda mundula*
Araneidae	*Gea sp.*
Salticidae	*Telmonia dimidiata*
Salticidae	*Bianor sp.*
Theridiidae	*Parasteatoda*
Oxyopidae	*Oxyopes sp 1*
Araneidae	*Araneus*
Salticidae	*Telamonia dimidiata*
Oxyopidae	*Oxyopes javanus*
FAMILY	SPECIES
Araneidae	*Neoscona sp.*
Lycosidae	*Pardosa shyamae*
Oxyopidae	*Hamataliwa sp.*
Salticidae	*Plexippus sp.*
Oxyopidae	*Oxyopes shweta*

Continued on next page.

Salticidae	*Rhene*
Linyphiidae	*Neriene radiata*
Theridiidae	*Parasteatoda mundula*
Theridiidae	*Argyrodes sp.*
Salticidae	*Bianor sp.*
Theridiidae	*Theridion sp.*
Lyniphyidae	*Neriene radiata*
Salticidae	*Rhene flavigera*
Araneidae	*Eriovixia laxaceles*
Salticidae	*Myrmarachne*
Lycosidae	*Lycosa sp.*
Salticidae	*Brettus anchorum*
Thomisidae	*Thomisus lobosus*
FAMILY	SPECIES
Tetragnathidae	*Leucauge sp.*
Lycosidae	*Pardosa sp. 2*
Oxyopidae	*Oxyopes javanus*
Salticidae	*Plexippus sp.*
Araneidae	*Neoscona sp*
Oxyopidae	*Oxyopesshweta*
Araneidae	*Neoscona mukerjei*
Araneidae	*Neoscona theisi*
Oxyopidae	*Oxyopes javanus*
Gnaphosidae	*Herphyllus*
Salticidae	*Hyllus semicupreus*
Pisauridae	*Nilus sp.*
Oxyopidae	*Oxyopes kusumae*

Lycosidae	*Pardosa songosa*
Araneidae	*Neoscona sp*
FAMILY	**SPECIES**
Tetragnathidae	*Leucauge sp.*
Lycosidae	*Lycosa sp.*
Oxyopidae	*Oxyopes sp 1*
Theridiidae	*Parasteatoda mundula*
Araneidae	*Aranaeus sp.*
Lycosidae	*Pardosa songosa*
Thomisidae	*Thomisus sp.*
Hersiliidae	*Hersilia sp*
Lycosidae	*Pardosa*
Eutichuridae	*Cheiracantheium sp*
Lycosidae	*Pardosa sp.*
Araneidae	*Neoscona mukerjei*
Lycosidae	*Pardosa sp. 1*
Theridiidae	*Euryopis sp.*
Lycosidae	*Pardosa shyam*
FAMILY	**SPECIES**
Salticidae	*Phintella vittata*
Salticidae	*Brettus anchorum*
Sparassidae	*Olios sp.*
Hersiliidae	*Hersilia savignyi*
Theridiidae	*Theridion sp.*
Salticidae	*Plexippus paykulli*
Theridiidae	*Parasteatoda sp.*
Thomisidae	*Mastira menoka*

Continued on next page.

Theridiidae	*Parastaetoda sp.*
Thomisidae	Species 1
Eutichuridae	*Cheiracanthium melanostomum*
Theridiidae	Species 2
Salticidae	*Evarcha sp.*
Theridiidae	*Theridion sp.*
Theridiidae	Species 3
Uloboridae	*Uloborus krishnae*
Theridiidae	Species 4
Salticidae	*Myrmarachne sp.*
Theridiidae	*Argyrodes sp.*
Uloboridae	*Zosis geniculata*
Theridiidae	*Parasteatoda sp.*
Theridiidae	*Parasteatoda mundula*
Eutichuridae	*Cheiracanthium sp*
FAMILY	SPECIES
Salticidae Insects	*Plexippus*
Mantis religiosa	—
Formicidae	—
Anisoptera	—
Coccinellidae	—
Reduviidae	—
Forficulidae	—
Nabidae	—
Cantharidae	—
Vespidae	—
Coenagrionidae	—

For farmers, the major concern besides care for the seed and the soil is keeping the crop free from pest attacks. Pest control ensures consistent production of the crop with appropriate weather and nutrients in the soil. However, as a consequence of the Green Revolution, farmers were exposed to the use of harmful chemicals, to which the pests kept becoming resistant. The indiscriminate production and synthetic chemicals destroy useful non-target insects and make the "pests" more resilient. This causes life-threatening consequences to the biodiversity in farmland, the surrounding natural areas, and humans and livestock.

The chemicals used in pesticides enter the food chain and affect every level through bioaccumulation. At Navdanya, we practice pest management by working with the biodiversity of plants and insects, not trying to exterminate them with pesticides and herbicides. These alternative biodiversity-based solutions are safe for crops, our bodies, and ecosystems. Biodiversity also contributes to food security because it protects pollinators, who contribute to one-third of the food we eat.

Biodiversity of Pollinators

Pollinators are biological vectors that assist in the transfer of the pollen grains from the anthers of the flower to the stigma (female reproductive organ of the flower). Thus helping in cross pollination. The next stage is fertilization and seed and fruit formation. Pollinators help in enabling fertilization in terrestrial flowering plants. A diverse group of mammals, birds, reptiles, and insects facilitate pollination in approximately 87.5% of wild and cultivated flowering plants. Insects, chiefly bees, flies, butterflies, and beetles, are the most important pollinators. They help in improving the crop production of seeds and fruits.

However, the use of chemicals and intensive agriculture, along with habitat loss, diseases, pathogen, etc., have led to pollinator decline worldwide. A diverse group of animals acts as pollinators. At Navdanya farm, we found 16 species of bees, three species of flies, one species of beetle, 21 species of butterflies, five species of birds, and two species of mammals that help in the pollination of 1/3 of the food crops grown at the farm.

Pollination is a key mechanism in promoting plant biodiversity in wild and human-managed ecosystems that depend on pollinators to ensure food security. The majority of non-timber forest products and crop production are benefits of pollinator services. The byproducts from wild and managed pollinating agents support the livelihoods of local communities. Pollinators, chiefly bees, aid in the reproduction of 87.5% of the world's flowering plants. Maintaining natural resources and dietary diversity necessitates active pollinator contributions. Animals from diverse taxonomic groups are pollinators of the majority of the flowering plants of the world. Over 18,000 bees, other insects, and diverse vertebrate pollinators pollinate plants. Pollination is a free-of-charge contribution to the wild and cultivated plants that cannot be replaced by any state-of-the-art technology. The current destruction of habitats, chemicals, the introduction of nonnative plants and pollinators, diseases, and pests are the driving forces of their decline.

At the Navdanya farm, we strive to maintain the natural balance that attracts pollinators and a plethora of other organisms that are beneficial to the crops and wild plants.

Table 3: Pollinators

GROUP	COMMON NAME	FAMILY	SCIENTIFIC NAME
Bees			
	Ground nesting	Andrenidae	*Andrena flavepis*
	Blue banded	Apidae	*Amegilla zonata*
	Rock	Apidae	*Apis dorsata*
	Dwarf	Apidae	*Apis florea*
	Asian honey	Apidae	*Apis indica*
	European	Apidae	*Apis mellifera*
	Bumble	Apidae	*Bombus haemorrhoidalis*
	Small Carpenter	Apidae	*Ceratina smaragdula*
	Stingless bee	Apidae	*Tetragonula irridipennis*
	Large Carpenter	Apidae	*Xylocopa aestuens*
	Mining bee	Halictidae	*Halictus spp.*
	Halictus sp. 1	Halictidae	*Lassioglossum spp.*

	Halictus sp. 2	Halictidae	*Nomia interstitialis*
	Cleptoparasitic bee	Halictidae	*Sphecoides spp.*
	Leaf Cutter bee	Mgachilidae	*Megachilae lanata*
	Orchid bee	Mgachilidae	*Osmia adae*
Flies			
	Syrphid Fly	Syrphidae	*Syrphid sp.*
	Drone Fly	Syrphidae	*Syrphid sp.*
	Tachinid fly	Tachinidae	*Tachinid sp.*
Beetles			
	Asian Lady Beetles	Coccinelidae	*Harmonia axyridis*
GROUP	COMMON NAME	FAMILY	SCIENTIFIC NAME
Butterflies			
	Cabbage White	Pieridae	*Pieris rapae*
	Painted Lady	Nymphalidae	*Venessa cardui*
	Lime Butterfly	Papilionidae	*Junonia lemon*
	Peacock Pansy	Nymphalidae	*Junonia almana*
	Common Grass Yellow	Pieridae	*Eurema hecabe*
	Plain Tiger	Nymphalidae	*Danaus chrysippus*
	Striped Tiger	Nymphalidae	*Danaus genutia*
	Plum Judy	Lycaenidae	*Abisaraecherius*
	Chocolate Pansy	Nymphalidae	*Junoniaiphita*
	Glassy Tiger	Nymphalidae	*Paranticaaglea melanoides*
	Common Sailor	Nymphalidae	*Neptishylas*
	Common Wonderer	Pieridae	*Pareronia valeria*
	Common Jezebel	Pieridae	*Delias eucharis*
	Crimson Rose	Papilionidae	*Pachliopta hector*
	Common Mormon	Papilionidae	*Papiliopolytes*
	Red Pierrot	Lycaenidae	*Talica danyseus*

Continued on next page.

Blue Tiger	Nymphalidae	*Tirumala limniace*
Dark Blue Tiger	Nymphalidae	*Tirumala septentrionis*
Common Jester	Nymphalidae	*Symbrenthia hippoclus*
Common Map	Nymphalidae	*Cyrestis thyodamas*
Tiny Grass Blue	Lycaenidae	*Zizula hylax*
Clouded Yellow	Pieridae	*Colia scroceus*

GROUP	COMMON NAME	FAMILY	SCIENTIFIC NAME
Birds			
	Rose-ringed Parakeet	Psittacidae	*Psittacula krameri*
	Plum-headed Parakeet	Psittacidae	*Psittacula cyanocephala*
	Crimson Sunbird	Nectariniini	*Aethopyga siparaja*
	Purple Sunbird	Nectariniini	*Netarinia asiatica*
	Oriental white-eye	Zosteropidea	*Zosterops palpebrosus*
Mammals			
	North. Palm Squirrel	Squiridae	*Funambulus pennantii*
	Indian Pipistrelle	Vespertilionidae	*Pipistrellus coromandra*

Birds are avid predators useful in pest control, seed dispersal, and pollination. One can find 71 different species of birds in the farm. A diversity of birds can be maintained by providing them with food sources and nesting habitats. Navdanya farm is not only an organic agroecosystem but also a tiny sanctuary in itself. In addition to crops, it has a huge diversity of trees shrubs and herbs—which is a paradise of biodiverse bird assemblages.

Table 4: Pollinators (birds)

GROUP	COMMON NAME	FAMILY	SCIENTIFIC NAME
Egrets			
	Little egret	Ardeidae	*Egrettagarzetta*
	Cattle egret	Ardeidae	*Bubulcus ibis*

Group	Common name	Family	Scientific name
Vultures, Kites			
	Black kite	Accipitridae	*Milvus migrans*
	Black-shouldered	Accipitridae	*Elanus caeruleus*
	Himalayan griffon	Accipitridae	*Gyps himalayensis*
	Cinereous vulture	Accipitridae	*Aegypius monachus*
GROUP	COMMON NAME	FAMILY	SCIENTIFIC NAME
Eagles			
	Crested serpent	Accipitridae	*Spilornis cheela*
	Steppe eagle	Accipitridae	*Aquila nipalensis*
	Imperial eagle	Accipitridae	*Aquila heliaca*
	Eurasian sparrowhawk	Accipitridae	*Accipiter nisus*
	Shikra	Accipitridae	*Accipiter badius*
Lapwings			
	Red-wattled	Charadriinae	*Vanellus indicus*
Pigeons, Doves			
	Rock Pigeon	Columbidae	*Columba livia*
	Spotted Dove	Columbidae	*Streptopelia chinensis*
	Oriental turtle dove	Psittacidae	*Streptopelia orientalis*
Parakeets			
	Rose-ringed	Psittacidae	*Psittacula krameri*
	Plum-headed	Psittacidae	*Psittacula cyanocephala*
Cuckoos			
	Pied cuckoo	Cuculidae	*Clamator jacobinus*
	Asian koel	Cuculidae	*Eudynamys scolopaceus*
Coucals			
	Greater coucal	Centropodidae	*Centropus sinensis*

Continued on next page.

Owls			
	Spotted owlet	Strigidae	*Athene brama*
Swifts			
	Asian palm swift	Apodidae	*Cypsiurus balasiensis*
	House swift	Apodidae	*Apus nipalensis*
GROUP	COMMON NAME	FAMILY	SCIENTIFIC NAME
Kingfishers			
	White-breasted	Alcedinidae	*Halcyon smyrnensis*
Bee-eaters			
	Green bee-eater	Meropidae	*Merops orientalis*
Rollers			
	Indian roller	Coracidae	*Coracias benghalensis*
Hoopoe			
	Hoopoe	Upupidae	*Upupa epops*
Hornbills			
	Indian grey hornbill	Bucerotidae	*Ocyceros birostris*
Barbets			
	Brown-headed barbet	Megalaimidae	*Megalaima zeylanica*
Woodpeckers			
	Black-rumped flameback	Picidae	*Dinopium benghalense*
Wagtails, Pipits			
	Grey wagtail	Motacillinae	*Motacilla cinerea*
	Yellow wagtail	Motacillinae	*Motacilla flava*
	White wagtail	Motacillinae	*Motacilla alba*
	Tawny Pipit	Motacillinae	*Anthus campestris*
Cuckoo shrikes			

	Large cuckoo shrike	Campephagidae	*Coracina macei*
Minivets			
	Small minivet	Campephagidae	*Pericrocotuscinnamomeus*
Bulbuls	Red-vented bulbul	Pycnonotidae	*Pycnonotus cafer*
	Black bulbul	Pycnonotidae	*Hypsipetes leucocephalus*

Table 5: Plants used for pest control on crops

NAME	BOTANICAL NAME	PART USED
Marigold	*Tegetes eracta*	flowers/leaves
Dainkan	*Meleaa azadaract*	Seed/leaves
Vitex	*Vitex nigundo*	Leaves
Pongam tree	*Punica granatum*	Rind
Datura	*Datura metal*	Leaves/fruit
Beshrum	*Ipomoea carnea*	Leaves
Garlic vine	*Mansoa alliacea*	Leaves
Arimesia spp.	*Artimesia annua*	Leaves/seed
Sullu	*Euphorbia royliana*	Exudates
Xanthoxylum	*Xanthoxylum aromaticum*	Leaves/fruits
Urtica	*Urtica dioica*	Above groundpart
Vasaca	*Adhatoda vasica*	Leaves
Hedychiaum	*Hedychium spicatum*	Tuber
Walnut	*Juglans regia*	Leaves/bark
Rumex	*Rumex nepalensis*	Leaves
Onion	*Allium cepa*	Bulb
Garlic	*Allium spp.*	Bulb
Camphor	*Cinnamum camphora*	Leaves
Sapium (Reetha)	*Sapindus mukkorosi*	Fruits

Biodiversity for Pest Control

Pest Management through Ecological Functions of Biodiversity:
The International Experience

Industrial agriculture is currently causing a massive decline in biological diversity, and this is considered one of the major contributors to the Anthropocene Extinction. It is the sixth major extinction event on our planet. The cause is clearly due to multiple human activities that are degrading our planet through habitat loss, pollution, climate change, and toxic chemicals. Industrial agriculture is clearly responsible for most of the habitat loss, a major contributor to climate change and major contributor to the decline of many species such as bees, birds, and frogs through toxic chemicals.

The United Nations Millennium Ecosystem Assessment Synthesis Report (MA Report 2005) raised serious questions about the sustainability of many of our current agriculture practices. "Over the past 50 years, humans have changed ecosystems more rapidly and extensively than in any comparable period of time in human history, largely to meet rapidly growing demands for food, fresh water, timber, fiber, and fuel. This has resulted in a substantial and largely irreversible loss in the diversity of life on Earth."

The IAASTD report concluded that "Business as usual was not an option" and strongly recommended the adoption of agricultural systems that used ecological systems that regenerated the environment rather than the current industrial, agricultural models that are severely degrading the environment (IAASTD 2008).

The best regenerative, organic farmers redesign farming systems so that they have a series of integrated systems that prevent pests and diseases from giving the crop a significant advantage. The aim is to have a whole systems approach that results in a resilient, low-input, high-output farm. This is where ecological sciences are applied to agriculture to produce systems based on agroecology.

5.16 *Eco Functional Intensification (EFI)*

Eco Functional Intensification (EFI) is the process of how a farm can move from being reliant on toxic and environmentally damaging inputs to a highly productive and resilient agroecological system that regenerates the environment.

EFI is defined as *an ecosystem-based regenerative production system that manages biodiversity to optimize ecosystem services. The aim is to maximize the multi-functional benefits of ecological functions rather than synthetic chemical intensification.*

Eco Functional Intensification optimizes the performance of ecosystem services. These services include pest and disease regulation, water holding and drainage, soil building, soil biology and fertility, nutrient cycling, nitrogen fixation, photosynthesis and carbon sequestration, a diverse agricultural crop and animal species, pollination, and many others.

The greater the biological complexity designed into a farming system, the less the chances for pests and pathogens to colonize and dominate that system. The aim is to create robust, sustainable, biodiverse systems with mechanisms that prevent and control most of the pest, disease, and weed problems and help increase the bioavailability of nutrients. These types of regenerative organic farming systems do exist and require minimum of input costs, making them the most efficient in returns to the farmer and the environment.

Soil health is the key principle to successful regenerative farming. Correctly balanced soil ensures minimal disease and insect damage. There is a large body of good scientific evidence showing that plants growing in fertile soils are more resistant to pests and diseases than plants that are deficient or stressed due to poor soils and or poor management.

An increasing number of scientific studies are showing that healthy plants produce a range of compounds that prevent or reduce damage from pests and diseases, particularly phenolic and flavonoid antioxidants. Interestingly, other research shows that these protective compounds protect their host plants, but they are also beneficial to the health of people who consume them. These compounds have been shown to have multiple benefits, such as being an anti-inflammatory for reducing the pain of rheumatism, arthritis, headaches, asthma and heart disease, and anti-cancer properties. Several studies show that organic foods have higher levels of these types of beneficial phytonutrients.

There is a growing body of evidence showing that healthy plants send out scent signals to each other warning of disease and insect attacks and that these plants will then generate a range of protective compounds to prevent damage. Researchers are currently studying a range of compounds plants emit when under pest attack that attracts beneficial predators to control the pests.

244 Agroecology and Regenerative Agriculture

Many years of research have shown that well-balanced soils with high levels of calcium, humus, and a neutral pH encourage a range of beneficial species and suppress pests and diseases.

These soils are rich in beneficial organisms like Trichoderma that control pathogens such as Rhizoctonia, Phytophthora, and Amilleria. Actinomycetes control many pests and diseases. Predatory nematodes control root burrowing nematodes, and organisms such as Metarhizium and Bacillus thuringiensis kill a range of insects.

Applying Eco-Functional Intensification (EFI)

The most efficient method of dealing with pests and diseases is to be proactive and have a pest management plan. Generally, the best results are obtained by developing a plan that uses a range of strategies taking a whole-farm approach.

Unfortunately, in most agricultural systems, pest management is an ad hoc process. It is either a belated reaction to a pest event or a very inefficient spray program that usually kills all the beneficials, and cause environmental damage and health problems.

Integrated Pest Management (IPM) has been introduced to many industries and is seen as a useful starting point in moving towards an agroecological system. The IPM tools of monitoring, setting pest level thresholds, and 'Hot Spot' spraying are very useful.

Effective monitoring is not exclusive to IPM and has always been regarded as an essential tool in good farming. There is an old saying: *"The footsteps of the farmer are the best fertilizer."*

This saying refers to the fact that monitoring and understanding what is happening in the crop and the farm as a whole is one of the most important management tools as it allows the farmer to take timely actions to prevent crop damage and loss.

Good regenerative organic farmers move beyond IPM by applying Eco Functional Intensification. One of the great advantages of Eco Functional Intensification is that once these systems are in place the ecology is doing the work to control the pests and diseases without the help of the farmer.

This means that the pests and diseases should be continuously controlled by the ecological systems the majority of the time. However, no system, natural or manmade is infallible. Good farmers will monitor and have a back-up strategy to deal with problems when they arise.

Biological methods of controlling pests are excellent examples of Eco Function Intensification. A range of ecological solutions are used to replace the need for spraying to kill the pests and diseases. The ecology does the work.

5.17 *Insectaries: Beneficial Insects and Their Host Plants*

Insectaries are groups of plants that attract and host beneficial insects, arthropods, and higher animal species. These are the species that remove arthropod (insect) pests from farms, orchards, and gardens. They are known collectively as beneficial.

Many beneficial insects have a range of host plants. Some useful species such as parasitic wasps, Hoverflies, and Lacewings have carnivorous larvae that eat pests; however, the adult stages live mostly on nectar and pollen from flowers. Flowers provide beneficial insects with concentrated forms of food (pollen and nectar) to increase their chances of surviving, immigrating, and staying in the area. Very importantly, flowers also provide mating sites for beneficial insects, allowing them to increase in numbers.

Without these flowers on a farm, the beneficial species die and do not reproduce. Most farming systems eliminate these types of plants as weeds, so consequently, they do not have enough beneficial insects to get good pest control. Buying and releasing commercial quantities of these insects is usually very expensive, especially if they cannot reproduce due to lack of suitable food.

Parasitic wasps prefer very small flowers; however, they have been found in large scented flowers such as water lilies. Flowers with high nectar and pollen content are regarded as the most valuable. Many weed species have these characteristics and are therefore very important in the control of insect pests.

Research into insectaries has been conducted at the University of California, Davis, the Dietrich Institute in California, Michael Field Research Station, Wisconsin, Rutgers University New Jersey, Lincoln University in NZ, FiBL in Switzerland, and several European universities. They have shown that planting these host plant species as ground covers, in rows, or in marginal areas, can cause a dramatic decline in pest species.

Farmers in the USA who have planted out rows of these host plants, as 'insectaries', in their fields no longer have to spray and have similar levels of pest control as their neighbors who are heavily spraying with toxic chemicals.

Encouraging nectar and pollen-rich flowers in and around the farm will improve the efficiency of these areas by changing the species mix in favor of these benefits. Ongoing research is determining the most effective mixes of plant species and distances between these nature strips.

Tansyleaf, a North American annual, is a good pollen source for adult hoverflies. A New Zealand study showed that aphid numbers were less than half the normal levels in cabbages and canola fields surrounded by a half-meter band of tansyleaf.

Replicated trials of eggplants and other vegetables conducted at Rutgers University, New Jersey, have shown that flowering dill, coriander, and fennel make good insectary plants for parasitic wasps.

University of California Davis researchers have shown that high levels of vegetation species diversity will ensure a constant low population of many arthropods that serve as "food" for the beneficials. The vegetation also helps to protect the beneficials and will ensure that they will stay in the area.

Taller host vegetation will contain significantly more beneficial insects than short vegetation. It is similar to high-rise buildings holding more people than single-story houses. The research also showed that a high diversity of host plant species resulted in higher levels of beneficial insects and better control of pest species.

In an experiment to produce effective insectaries, UK and New Zealand scientists made ridges about 2 m wide and 500 mm high by two-way plowing. They were sown with cocksfoot, Yorkshire fog, perennial ryegrass, and creeping bent. The first two formed tussocks, the others formed mats. The tussock grasses are fast growers and competitive, so weeds have not been a problem. After two years, there were 1,500 predators per square meter, roughly 10 times the density of good quality hedgerows. After about five years, trees and shrubs have begun to grow similar to hedgerows.

In Europe, NZ, Australia, and the US, farmers can get lists of insectary plants to introduce to their farms. Unfortunately, very little research has been conducted on designing insectaries for other climate conditions, especially for the tropics.

Three Rules for Designing Insectaries

1. Any flowering plant that attracts bees is suitable as an insectary plant.
   Beneficial insects prefer species that are rich in pollen and nectar.
2. Smaller flowers are best for parasitic wasps.

3. The greater the diversity of species, the more effective the insectary system.

Research from Lincoln University in NZ and in California has shown that it is best to weed in stages, always leaving good refuges of weeds in and around the farm to ensure a healthy supply of beneficial species. Never control all the weeds on the farm at the same time.

Dust interferes with predatory insects' ability to locate hosts and can lead to outbreaks of pests like spider mites. Planting insectary plants as windbreaks and ground covers will reduce dust.

Studies in the USA have shown that the value of the insects living and breeding in insectaries can be as much as US$30,000 per acre (US$75,000 per hectare) if these insects were purchased from commercial suppliers.

Multiple Examples of Biological Control Methods
Trap Crops

Trap Crops are a variation of insectaries and are used to trap pest species. There is a range of methods and types of crops that are used.

1. **Continuous Preferred Hosts**

   These work by drawing the pest species away from the crop because they prefer the trap crop to the cash crop. American cotton farmers plant alfalfa (lucerne) rows in their fields because Lygus Bugs prefer alfalfa over cotton. The farmers alternately mow half the row for the full length every two weeks. This creates a continual strip of alfalfa that is in the correct state for the Lygus Bugs as well as leaving most of the benefits in the alfalfa.

2. **Timed Alternate Hosts**

   These work by planting crops that attract the pest species before or after the season. The pests are then destroyed to break the breeding cycles and reduce the pest population.

   Examples of these are crops that attract nematodes. These are usually planted early in the season and plowed in as green manure before the nematodes begin laying eggs. If used properly, this system will break the pest cycle, reduce weeds, provide valuable organic matter, and slowly release nutrients for the cash crop.

   A variation is to plant the trap crop straight after the cash crop. Usually, the pest species is at its greatest at this point. A combination of a trap crop and a rotation cash crop the following year has been shown to be the most effective in significantly controlling pests.

Another version is to plant a few small areas of the cash crop a few weeks earlier and plow it is just before planting the main cash crop. Timed properly, this can significantly reduce pests. In the US, potato cyst nematodes are reduced by 80% by plowing out the trap potato crop before the nematodes have time to reproduce.

Some American cotton farmers plant narrow strips of cotton as a border around their fields several weeks before planting the cash crops. Boll weevils and other pests congregate in the trap cotton and are controlled by spraying an insecticide. This reduces the number of pest insects in the cash crop.

The use of alternate hosts that attract the pest early in the season can be useful. An examples of this is the use of Jaboticabas trees flowering just before lychees or mangos. These will attract the monolepta beetles where they can be destroyed before they attack lychee and mango flowers.

Australian research has shown that chickpeas make the best trap crop for the pesticide-resistant Heliothis (*Helicoverpa armigera*). Linseed, canola and field peas were shown to be good trap crops for Heliothis and host plants for predators and parasitoids such as lacewings, ladybeetles, and wasps.

3. **Lures**

Insectaries can be used as trap crops by placing/spraying lures and baits to attract the pest species out of the cash crop and into the predator-rich insectary.

Repellent Species

Some plants repel insect pests. Interplanting repellent species within the crop makes it less attractive to the pest. Having a non-crop plant that is a preferred host planted near the crop will attract the pest away from the crop.

Push-Pull Method

The best systems work by integrating several of the bio-control strategies into a whole systems approach.

The push-pull method in corn is an excellent example of an organic method that integrates several of these elements to achieve substantial increases in yields. This is significant because corn is the key food staple in

Africa and Latin America. The push-pull system was developed by scientists in Kenya at the International Centre of Insect Physiology and Ecology (ICIPE), Rothamsted Research, UK, and with the collaboration of other partners.

The push-pull method is an excellent example of Eco Function Intensification as an integrated production system. It uses the combination of a cover crop and a trap crop to prevent stem borers and the Striga parasite in corn.

Desmodium is planted to repel the stem borer and also to attract the natural enemies of the pest. Its root exudates stop the growth of Striga, which is a parasitic weed of corn. Napier grass is planted outside of the field as a trap crop for the stem borer. The desmodium repels (push) the pests from the maize, and the Napier grass attracts (pull) the stem borers out of the field to lay their eggs in it instead of the corn. The sharp silica hairs on the Napier grass kill the stem borer larvae when they hatch to break the life cycle and reduce pest numbers.

High yields are not the only benefits. The system does not need synthetic nitrogen as desmodium is a legume and fixes nitrogen. Soil erosion is prevented due to a permanent ground cover. Very significantly, the system provides quality fodder for stock. Napier grass and desmodium are systematically harvested to provide fresh fodder for livestock. Livestock can also graze down the field after the maize is harvested. Many push-pull farmers will integrate a dairy cow into the system and sell the surplus milk as a regular source of income.

Figure 2: The desmodium suppresses weeds, adds nitrogen, conserves the soil, repels pests, and provides high protein stock feed.

Barriers

Trap crops and permanent insectaries can be used as a barrier to prevent the entry of pests into a cash crop. Farmers in the USA have planted several rows of tall crops or grass-like wheat, corn, rye, and vetch around fields of vegetables. These create a barrier that slows down the entry of aphids that transmit viruses to crops.

Farmers in Myanmar plant barriers of sunflowers around their fields. These work as barriers to stop pests from entering as they get attacked by the beneficials in the sunflower barriers. The beneficials can also enter the field to protect the crop.

Many farmers, traditionally, have hedge borders of different plants along the pathways and farm boundaries with a diverse collection of native and introduced species. This is very common in some parts of Africa. These border hedges act as refuges for beneficials as well as barriers for pests.

Figure 3: Hedge borders along pathways in Kenya

Higher Animals

Insectaries or nature strips are also hosts to valuable higher animals. A large range of higher species plays a very significant role in controlling pests in agriculture.

Many bird species will eat pest insects. Examinations of the stomach contents of most bird species commonly found on farms show that they can consume large numbers of insects. Each bird can eat thousands of insects per year.

Other published research has shown that one of the major problems with widespread herbicide use is the loss of the habitat refuges of birds and beneficial insects. The bird numbers plummet in these districts, resulting in higher pesticide use.

Studies of orchards using total exclusion netting have demonstrated that these orchards need to use more insecticides to kill pests that birds previously ate.

Dense bushes, small trees, shrubs, and bamboos are the plants to use to attract insect-eating birds. Most of these are small birds and like to shelter and nest under thick canopies to avoid predators. Microbats are very effective in controlling many of the night-flying insect pests. Each bat has to eat one-third of its body weight in insects every night. This is a massive number of insects. In Europe and the USA, some farmers build bat houses to keep them on-farm. They also place lights in the crop to attract insects for the bats to eat. Many farmers also collect the bat guano around the houses as fertilizer.

Microbats can be used to control fruit sucking and piercing moths by putting a battery-powered light in the sections of the orchards where pest control is needed. The bats are attracted to light, as they know light attracts many insects. They can locate and eat nearby moths with their sonar.

Lizards, frogs, and toads eat a wide range of pest insect species. A light placed on the ground at night will attract frogs and toads to consume the insects that are drawn by the lights. Many pest beetle and moth species can be controlled this way.

Poultry (chickens, ducks, peafowl, and guinea fowl) are very effective in cleaning up pest species like grasshoppers and beetles. These animals have been used traditionally in all farming cultures as an essential part of pest control.

Owl nesting boxes in high trees are proving effective in controlling rats. It is important to have perch trees and a cleared border, at least two meters wide around the field, so that the owls can see and catch the rats as they run in and out of the fields.

Planting Non-Pest Host Species and Pest-Resistant Varieties

Where possible, it is important to source crop varieties that are resistant to major pests and diseases. Some weed or garden plants can host insect pests. As an example, some pest beetle species such as cane beetle, monolepta, and rhyparida larrvae live on the roots of grasses.

Many pest rat species nest and live in long grasses. Replacing these with shrubs and trees, and other flowering plants will reduce their numbers. Along with the introduction of beneficial plant species, these measures will significantly reduce the damage the pests cause in crops.

Purchasing Beneficial Arthropods

Many beneficial insects can now be purchased. The following groups of arthropods are usually available:
+ Predatory nematodes
+ Predatory mites
+ Trichogramma, telenomus and other parasitic wasps
+ Lacewings
+ Lady Beetles
+ Assassin Bugs and other predatory bugs

Baits, Lures, Traps, and Pheromone Disruptors

A range of traps, baits, and lures are used to control insects. These are some of the best methods as they concentrate on controlling pest species without adversely affecting non-target species.

Examples of these are protein hydrolysate baits for fruit flies. These tend to mostly attract females; however, they will also attract many males. The flies feed on enterobacteria that live in the protein bait. The bait should be contained in a vessel that prevents escape or has enough water to drown them.

Pheromones or parahormones can be used as baits or as methods to disrupt mating. Variations of these are now being used very effectively with codling moths and fruit flies and are available commercially.

Borax and sugar baits can be used for the control of a large range of insect pests, particularly cockroaches, termites, and ants. Use soil tests to ensure that the levels of boron are not too high as it can be toxic to plants. If the tests show boron deficiencies, then it can be used without causing any soil problems.

Sticky pastes can be applied around the trunks of trees to trap insects, such as ants that climb up the trunk. Colored sticky traps are used; however, they tend to be better as a method of monitoring the pest numbers rather than as a significant control method.

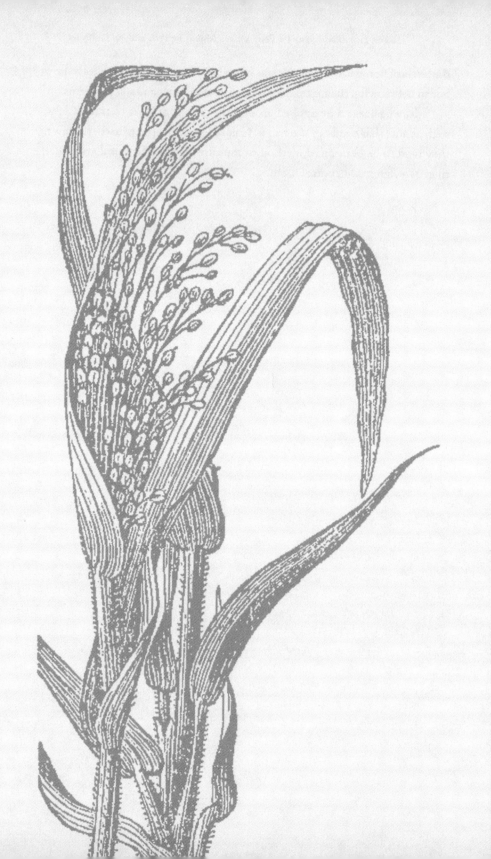

SECTION 6 Food, Nutrition, and Health

The industrial agriculture model has been promoted with the justification that it is the only answer to food security. However, globally, 1 billion people are structurally hungry. In India, every fourth Indian is hungry; every second child is malnourished.

Further, because of industrial agriculture and industrial food, 3 billion people suffer from chronic diseases (Annam, *Manifesto on Food for Health*, Navdanya International September 2018). Colin Todhunter writes:

"Indian government data indicates that cancer showed a 5% increase in prevalence between 2012 and 2014 with the number of new cases doubling between 1990 and 2013. The incidence of cancer for some major organs in India is the highest in the world."

The increase in the prevalence of diabetes is also worrying. By 2030, India's number of diabetes patients is likely to rise to 101 million (World Health Organization estimate). The figure doubled to 63 million in 2013 from 32 million in 2000. Over 8% of the adult male population in India has diabetes. The figure is 7% for women. According to the WHO, almost 76,000 men and 52,000 women in the 30-69 age group in India died due to diabetes in 2015.

A study in The Lancet from a couple of years ago found that India leads the world in underweight people. Some 102 million men and 101 million women are underweight, which makes the country home to over 40% of the global underweight population.

Contrast this with India's surge in obesity. In 1975, the country had 0.4 million obese men, or 1.3% of the global obese men's population. In 2014, it was in the fifth position globally with 9.8 million obese men, or 3.7% of the global obese men's population. Among women, India is globally ranked third, with 20 million obese women, or 5.3% of the global population.

According to India's 2015–16 National Family Health Survey, 38% of under-5s are stunted (height is significantly low for their age). The survey also stated that 21% of under-5s are significantly underweight for their height, a sign of recent acute hunger. The prevalence of underweight children in India is among the highest in the world; at the same time, the country is fast becoming diabetes and heart disease capital of the world."

India now has the second-highest cancer rates among women.

Producing nutritious food without chemicals and pesticides has emerged as a health imperative. And pesticide-free farming contributes to nutrition, food, and health security.

Over 30 years, Navdanya's practice and research have shown that we can produce more nutrition per acre through biodiversity and regenerate people's health.

6.1 *Seeds of Hope to Address Hunger and Malnutrition: Health Per Acre*

Organic biodiversity-based mixed cropping is the foundation of the concept of health per acre. It is a system of farming that increases nutrition produced per acre of farmland. A great amount of food, as well as a variety of food produced and consumed at the local level, helps inequitable distribution. The system promotes growing traditional local foods and, hence, promotes the consumption of such foods at the local level. The wide variety of local food items covers the entire profile of nutrients required by the human body. Organic mixed cropping methods maximize the nutrition produced per acre and, hence, help control inflation of food items. Another reason why such a cropping method would control food price is that food produced and consumed locally avoids the huge cost of transportation and storage usually included in the price consumer pays for food items. The population, at large, usually knows quite a lot about local food items and their health benefit. As a result, educating people, especially women, with the various aspects of health and nutrition becomes easier. Implementation of such knowledge also becomes easier as adaption, availability, and cost are not mutually exclusive but rather facilitate one another. The approach focuses more on the root cause of the problem of undernutrition rather than on the treatment of current cases of malnutrition. Treatment is just one aspect of solving crises. However, irrespective of how sophisticated our treatments are, undernutrition cannot be eradicated until we make an adequate quantity of a variety of food available to the target population sustainably.

Nutrition produced per acre gives insight into the impact that the organic mixed cropping method can have on the health of the population. Until now, we have focused primarily on the yield per acre.

Looking at agriculture and health in terms of yield per acre makes an important assumption that maximizing the yield of specific food items would solve the undernutrition crisis. However, a few food commodities produced in large quantities is not an answer to hunger and malnutrition. Most agricultural commodities go for biofuel and animal feed. The fraction used as food cannot ensure the diversity of nutrients needed for health. An ideal blend and balance of nutrients are not available from the handful of globally traded commodities. To ensure proper nutrition, we need dietary diversification, and to ensure dietary diversification, we need to diversify our farmlands. There is a huge discussion that tries to find the answer to the question - which farming practice can ensure food security- organic mixed cropping or chemical mono copping? The yield per acre of monocultures of specific food items, used as a measure of effectiveness, appeared to favor conventional mono-cropping. However, when we change the metric to nutrition produced per acre of farmland in the two compared farming systems, strikingly different results came out. Diversity produces more nutrition and health per acre. Chemical monocultures produce more nutritionally empty, toxic commodities that harm health both through nutritional deficiency and through the presence of toxins.

What needs to be pointed out is whether abundant production of rice, wheat, corn, or soybean would solve the crisis of undernutrition or abundant production of all the different nutrients would. Organic biodiversity-based mixed cropping is sustainable, time-tested, reasonable, intelligent, cost-effective, and an ecological solution to the problem of malnutrition in India. The evidence is clear, the more the biodiversity, the higher the nutrition per acre.

However, the low productivity of industrial agriculture in terms of its nutrition is hidden through reductionism. The industrial system is not compared to agroecological systems. The system is reduced to one crop and then one part of a crop which is a commodity.

Modern plant breeding concepts like HYVs reduce farming systems to individual crops and parts of crops.

Crop components of one system are then measured with crop components of one another. Since the Green Revolution strategy is aimed at increasing the output of a single component of a farm, at the cost of decreasing other components and increasing external inputs, such a partial comparison is, by definition, biased to make the new varieties 'high yielding' even when, at the systems level, they may not be.

Traditional farming systems are based on mixed and rotational crop-ping systems of cereals, pulses, and oilseeds—with diverse varieties of each crop—while the Green Revolution package is based on genetically uniform monocultures.

No realistic assessments are ever made of the yield of the diverse crop outputs in the mixed and rotational systems. Usually, the yield of a single crop like wheat or maize is singled out and compared to yields of new varieties.

Even if the yields of all the crops were included, it is difficult to convert a measure of pulse into an equivalent measure of wheat, for example, because in the diet and in the ecosystem, they have distinctive functions. The protein value of pulses and the calorie value of cereals are both essential for a bal-anced diet, but in different ways, and one cannot replace the other.

Similarly, the nitrogen-fixing capacity of pulses is an invisible ecological contribution to the yield of associated cereals.

The complex and diverse cropping systems which are based on Indigenous varieties are therefore not easy to compare to the simplified monocultures of HYV seeds. Such a comparison has to involve entire sys-tems and cannot be reduced to a comparison of a fragment of the farm sys-tem. In traditional farming systems, production has also involved maintaining the conditions of productivity.

The measurement of yields and productivity in the Green Revolution paradigm is divorced from seeing how the processes of increasing output affect the processes that sustain the condition for/agricultural production. While these reductionist categories of yield and productivity allow a higher measurement of partial yields, they exclude the measurement of the ecolog-ical destruction that affects future yields.

6.2 *Yield is Not Output: The Myth of More Food*

The most common argument for chemicals in food and genetic engineering is that they are the only way to feed people. However, an analysis of the trends and impacts of the Green Revolution and genetic engineering make it evi-dent that chemicals and genetic engineering in agriculture are a guarantee for creating scarcity and hence increasing food insecurity. Because it is evolving in the monoculture paradigm, which focuses on single functions of single species, it fails to take the yields of diverse species and diverse functions of species into account. In fact, genetic engineering can only displace and destroy the diverse foods that account for food security in diverse food cultures.

The argument of increased food availability through industrial breeding, including genetic engineering, is illusionary on four counts.

1. Industrial breeding both in genetic engineering and the Green Revolution focuses on partial aspects of single crops rather than total system yields of multiple crops and integrated systems.

2. Industrial breeding focuses on yields of one or two global commodities, not on the diverse crops that people eat. Industrial breeding focuses on quantity per acre rather than nutrition per acre. In fact, nutrition per acre has come down as a result of industrial agriculture.

3. Industrial breeding, including genetic engineering, uses natural resources intensively and wastefully. If productivity is defined on the basis of resource use, industrial agriculture has very low productivity, and it undermines food security by using up resources that, if not wasted in a non-sustainable system of production, could have been directly used to produce more food.

4. Ecological alternatives can increase food supply through biodiversity intensification instead of chemical intensification and genetic engineering.

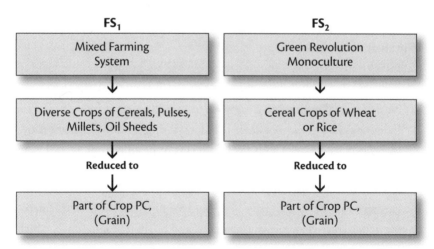

Figure 1: Green Revolution

6.3 *Changing the Metric From Yield Per Acre to Nutrition Per Acre*

Chemical agriculture has been based on the metric of "yield per acre," which is not a reliable measure of the productivity of an ecosystem. The central myth that has led to the displacement of diverse farmers' varieties by supposed high yielding varieties (HYVs) is that the former are low yielding, and the latter are high yielding and have higher productivity.

1. HYVs are not intrinsically high yielding. They merely respond well to chemicals and are more appropriately called High Respond Varieties (HRVs)

2. HRVs demonstrate a high partial yield because such varieties have been bred only to yield enhanced grain production with high chemical inputs. This increase in production of grain for the market is achieved by reducing the biomass for internal use on the farm, both for fodder as well as for fertilizer.

HRVs exhibit low total system productivity. In countries like India, the quantity of straw obtainable is important as fodder for livestock but HRVs fail to produce enough straw that is adequate in quality or quantity. The increase in marketable output of grain has been achieved at the cost of decrease of biomass for animals and soils and the decrease of ecosystem productivity due to overuse of resources.

3. Indigenous varieties often outperform HRVs in total system yield in the realistic conditions of the fields of small farmers. When the total biomass is taken into account, traditional farming systems based on Indigenous varieties are not found to be low yielding at all. In fact, many native varieties have higher yields both in terms of grain output as well as in terms of total biomass output (grain and straw) than the supposed HYVs that have been introduced in their place.

4. Farmers' varieties which have been bred for nutrition, have more micronutrients, trace elements, antioxidants, and phenolic compounds, which are vital to health. Food is for nourishment, and the relevant metric is the nutritional quality, not the weight of nutritionally empty commodities.

6.4 *Towards a Biodiversity-Based Productivity Framework*

According to the dominant paradigm of production, diversity goes against productivity, which creates an imperative for uniformity and monocultures. This has generated the paradoxical situation in which modern plant improvement has been based on the destruction of the biodiversity which it uses as raw material. The irony of plant and animal breeding is that it destroys the very building blocks on which the technology depends. Forestry development schemes introduce monocultures of industrial species such as eucalyptus, and push into extinction the diversity of local species which fulfils local needs. Agricultural modernization schemes introduce new and uniform crops into farmers' fields and destroy the diversity of local varieties.

The modernization of animal husbandry destroys diverse breeds and introduces factory farming.

This strategy of basing productivity increase on the destruction of diversity is dangerous and unnecessary. Monocultures are ecologically and socially non-sustainable because they destroy both nature's economy and people's economy.

In agriculture, forestry, fisheries, and animal husbandry, production is being incessantly pushed in the direction of diversity destruction. Production based on uniformity thus becomes the primary threat to biodiversity conservation and to sustainability, both in its natural resource and its socio-economic dimensions.

Not until diversity is made the logic of production, can diversity be conserved. If production continues to be based on the logic of uniformity and homogenization, uniformity will continue to displace diversity. 'Improvement' from the corporate viewpoint, or from the viewpoint of Western agricultural or forestry research, is often a loss for the Third World, especially for the poor in the Third World. There is therefore no inevitability that production should act against diversity. Uniformity as a pattern of production becomes inevitable only in a context of control and profitability.

Plant improvement in agriculture has been based on the 'enhancement' of the yield of desired product at the expense of unwanted plant parts. The 'desired' product is however not the same for agribusiness and Third World peasants. Which parts of a farming system will be treated as 'unwanted' depends on what class and what gender one belongs to. What is unwanted for agribusiness may be wanted by the poor, and by squeezing out those aspects of biodiversity, agriculture 'development' fosters poverty and ecological decline.

Overall productivity and sustainability is much higher in mixed systems of farming and forestry which produce diverse outputs.

Productivity of monocultures is low in the context of diverse outputs and needs. It is high only in the restricted context of output of 'part of a part' of the forest and farm biomass. "High yield" plantations pick one tree species among thousands, for yields of one part of the tree (e.g. pulpwood). "High yield" Green Revolution cropping patterns pick one crop among hundreds, such as wheat for yields of one part of the wheat plant (only grain).

These high partial yields do not translate into high total (including diverse) yields. Productivity is therefore different depending on whether it is measured in a framework of diversity or uniformity.

In the context of climate change, the relevant metric is climate resilience. The so called HYV's are low yielding in droughts, floods, and cyclones and are vulnerable to crop failure. Their yield falls to zero. Traditional varieties bred for salt resistance, flood resistance, and drought resistance have climate resilience in our times of climate change.

Diversity has been destroyed in agriculture on the assumption that it is associated with low productivity. This is however, a false assumption both at the level of individual crops as well as at the level of farming systems. Diverse native varieties are often as high-yielding or more high-yielding than industrially bred varieties. In addition, diversity in farming system has higher output at the total systems level than one-dimensional monocultures.

Comparative yields of native and Green Revolution varieties in farmer's fields have been assessed by Navdanya, a National Seed & Biodiversity Conservation Program and organic agriculture movement. Green Revolution varieties are not higher yielding under the conditions of low capital availability and fragile ecosystems. Farmers' varieties are not intrinsically low-yielding and Green Revolution varieties or industrial varieties are not intrinsically high-yielding. As Yegna Narayan Aiyer reports,

"The possibility of obtaining phenomenal and almost unbelievably high yields of paddy in India has been established as the result of the crop competitions organized by the Central Government and conducted in all states. Thus even the lowest yield in these competitions has been about

5300 Ibs/acre, 6200 Ibs/ acre in West Bengal, 6100, 7950, and 8258 Ibs/ acre in Thirunelveli, 6368 and 7666kg/ha in South Arcot, 11,000 Ibs/ acre in Coorg and 12,000 Ibs/acre in Salem".

The measurement of yield and productivity in the Green Revolution as well as in the genetic engineering paradigm is divorced from seeing how the processes of increasing single species, functions, and outputs affect the processes that sustain the condition for agricultural production, both by reducing species and functional diversity of farming systems as well as by replacing internal inputs provided by biodiversity with hazardous agrichemicals. While these reductionist categories of yield and productivity allow a higher measurement of harvestable yields of single commodities, they exclude the measurement of the ecological destruction that affects future yields and the destruction of diverse outputs from biodiversity-rich systems.

Productivity in traditional farming practices and in agroecological systems has always been high if it is remembered that very little external inputs are required.

Industrial agriculture and industrial processing are root causes of the non-communicable disease epidemic. Thinking of the soil as an empty container, plants as machines that run on chemical fertilizers as fuel, pests as "enemies" to be exterminated, food as matter that we stuff ourselves with, and our bodies as machines that run on food as fuel and need fixing externally when they break down, is at the root of the multiple crises of agriculture, food, and health we face. This mechanistic industrial paradigm does not have the epistemic or intellectual potential to understand the roots of the disease epidemic it has created nor does it offer lasting solutions to the problems of malnutrition and chronic diseases it has created.

The destruction of the Earth and her diversity, the destruction of the diverse cultures and rich knowledge systems by the mechanical mind, has left the Earth and humanity impoverished, and more ignorant in terms of living systems and living processes.

We need to make a paradigm shift from the violence of mechanistic reductionism inherent to industrial agriculture, industrial food and industrial medicine to the biodiversity-centered, ecological, and nutritionally sensitive paradigms of agriculture, food, nutrition, and health if we have to make a transition from being a malnourished and sick nation to a healthy nation.

Healthy soils are full of biodiversity. 1 gm of soil organic soil contains 30,000 protozoa, 50,000 algae, 400,000 fungi. One teaspoon of living soil contains 1 billion bacteria, which translates to 1 ton per acre. One square cubic meter of soil contains 1,000 earth worms, 50,000 insects, and 12 trillion roundworms.

Humus, which is the Latin word for living soil, is also the root of "human." We are connected to the soil. When soils are heathy, societies are heathy. When soils are sick and desertified, societies become sick.

Desertification of the soil is related to not returning organic matter to the soil. Soils rich in humus can hold 90% of its weight in water. Living soils are the biggest reservoir of both water and nourishment.

Healthy soils produce healthy plants. When the soil is healthy, with a diversity of living organisms, it is able to produce all the nourishment it needs, and all the nourishment plants need.

In nutrition sensitive agriculture we need to maximize nutrition per acre, not merely yield of nutritionally empty commodities per acre.

Industrial agriculture has been based on the false claim that it produces more food. However only 30% of the food we eat comes from large industrial plants. The rest comes from small farms. Only 10% of the corn and soya grown in the world is eaten by people as food. 90% goes to biofuel and animal feed.

Industrial agriculture is not a food system for nourishing the planet and people's health. It is a commodity system, producing profits for a handful of corporations, making both the Earth and humans sick.

The mechanistic, reductionist paradigm of chemical /industrial agriculture has reduced agriculture to the production of food as a commodity to be traded. For trade, what matters is mass and quantity, not quality. Food is getting degraded in quality, real food is disappearing and being replaced by nutritionally empty commodities, full of toxins, which cause harm to health.

Agriculture exclusively focused on selling agrochemicals as inputs for commodity production has reduced the measure of productivity of agriculture to "Yield per Acre". But yield per acre leaves out the most important aspects of food and farming that a nutrition sensitive agriculture needs. "Yield" measures mass, the quantity of a commodity, not the nourishing quality of food. Hence it is inadequate as a measure of food in the context of health and nutrition. Nor does "yield" measure the destruction of biodiversity that provides nourishment and health. "Yield" alone does not measure the high financial costs of toxic inputs which are trapping farmers in debt and pushing them to suicide. Nor does it measure the cost of the disease burden due to toxins in our food. Finally, yield per acre does not measure the ecological cost of chemical monocultures.

We need to move away from measuring "yield per acre" of nutritionally empty, toxic monocultures produced at high cost to measuring nutrition per acre of a diversity of crops.

Navdanya has been promoting biodiversity intensive nutrition sensitive agriculture for three decades in different states of India. In diverse ecosystems and diverse farming systems, diversity always systematically produces more nutrition per acre than monocultures. As an example, Navdanya's surveys carried out in Sikkim (which was recently declared a 100% organic state in 2016 by the PM Narendra Modi), confirms that biodiverse organic farms produce more nutrition per acre than chemical monoculture farms.

Nutrition sensitive agriculture produces more nutrition for people and the soil.

On the Navdanya farm, organic matter has increased up to 99%, zinc has increased by 14%, and magnesium has increased by 14%. We did not add these as external inputs. They have been produced by the billions of soil microorganisms that are in living soils. Healthy soils produce healthy plants. Healthy plants are then able to nourish humans. On the other hand, chemical farming has led to decline in soil nutrients, which translate into a decline in the nutrition content of our food.

Table 1: Nutrition produced per acre of farmland, Sikkim

NUTRIENT	ORGANIC MIXED CROPPING	CONVENTIONAL MONO-CROPPING
Protein	64.2 kg	55.5 kg
Carbohydrate	3–4.0 kg	331.0 kg
Fat	17.2 kg	18.0 kg
Energy	1,622,000 cals	1,710,000 cals
Carotene	3,154 mg	450 mg
Thiamine	2,330 mg	2,100 mg

Ribofavin	460 mg	500 mg
Niacin	9,800 mg	9,000 mg
B6	—	—
Folic Acid	80 mg	100 mg
Vitamin C	81,000 mg	0 mg
Choline	166,000 mg	0 mg
Calcium	305 g	50 g
Iron	29.3 g	11.5 g
Phosphorous	1,740 g	1,740 g
Magnesium	626 g	695 g
Sodium	145.2 g	79.5 g
Potassium	1,878 g	1,430 g
Chlorine	172 g	165 g
Copper	6,420 mg	2,050 mg
Manganese	3,030 mg	2,400 mg
Molybdenum	790 mg	190 mg
Zinc	14,240 mg	14000 mg
Chromium	48 mg	20 mg
Sulphur	645,000 mg	570,000 mg

6.5 *The Ayurvedic Approach to Health*

The combination of the depletion of biodiversity and nutrients in our food is translating into disease. Ayurveda as a science that recognized that the digestive system is central to our health. Even Western Science is beginning to realize what Ayurveda understood 5,000 years ago—that the body is not a machine, and food is not fuel that runs the machine on Newtonian laws of mass and motion. Food is not "mass;" it is living; it is the source of life and the source of health.

There is an intimate connection between the soils, the plants, our gut, and our brain. Our gut in a microbiome contains trillions of bacteria. To function in a healthy way, the gut microbiome needs a diverse diet, and a diverse diet needs diversity in our fields and gardens. A loss of diversity in our diet creates ill-health. Chemicals in agriculture kill soil microbes and the beneficial microbes in our gut.

There are more than 100 trillion microbes of 100 bacterial species in our gut. The gut microbiota contains 7 million genes or up to 360 bacterial genes for every human gene. Less than 10% of the genes in our body are human. There are 100,000 times more microbes in our gut than people on the planet.

The gut is increasingly being referred to as the second brain. It has its own nervous system, which is referred to as the enteric nervous system, or ENS, with 50–100 million nerve cells. Our bodies are intelligent organisms. Intelligence is not localized in the brain. It is distributed. And the intelligence in the soil, plants, and our bodies makes for health and well-being.

Loss of biodiversity in our fields and our diet over the last half-century with the spread of the industrial agriculture is not just leading to an ecological crisis. It is leading to a nutritional crisis and disease epidemic.

Ayurvedic science identifies *Agni* in the digestive tract as the great transformer. It recommends that we eat foods with a diversity of six tastes to have a balanced diet: sweet, sour, salt, pungent, hot, bitter, astringent.

Behind each taste are potentials for processes that create and sustain our body's self-regulating systems, creating emergent properties. Taste receptors do not just lie in the tongue but are distributed throughout the gastrointestinal tract and are located on sensory nerve endings and hormone-containing transducer cells in the gut wall.

New biological science is now finding out that the gut has sensors for different tastes, and different metabolic processes are governed by the diversity of tastes. As the research by Dr. Eric Seralini shows, the sophisticated intelligence in the complex ecosystem of our gut communicates with the food we eat. When we eat fresh and organic food, the regulatory processes that ensure health are strengthened. When we eat chemical food with toxins, the communication leads to disease.

Specific molecules and phytochemicals found in herbs and spices activate specific taste receptors and trigger particular metabolic processes. Sweet receptors stimulate glucose absorption into the bloodstream and release insulin from the pancreas when they sense glucose. As Mayer states, "The multitude of phytochemicals derived from a diet rich in diverse plants combined with the array perfectly matching sensory mechanisms in our gut, synchronizes our internal ecosystem, our gut micro biome with the world around us .... (pg 59 Mayer)."

Nutrition-sensitive agriculture recognizes that everything is food and everything is something else's food—from the soil, through the plants, and to our gut. Agriculture is not sensitive to the nutrition cycle and the food web that connects the Earth and our bodies; it desertifies the soil and our gut.

The sophisticated understanding of nutrition in Ayurveda has shaped the diversity and healthy basis of India's traditional food cultures. Today this rich nutrition sensitive agriculture and dietary culture is threatened with industrial monocultures, the invasion of junk food, industrially processed food grown with toxic chemicals, and fake food.

Both the health of the planet and our health has suffered. 75% of the planetary crisis of water, soils, biodiversity has its roots in industrial agriculture. Across India there is a water crisis because the so-called Green Revolution based on chemicals is extremely water intensive and has led to the diversion of rivers and mining of groundwater for intensive irrigation. More than 75% of the water resources have been destroyed and polluted due to chemical farming.

Chemicals have desertified our soils by failing to return organic matter that creates living soils. Our rich biodiversity of plants and food crops has been dramatically eroded with the spread of monocultures of nutritionally insensitive agriculture.

We evolved 200,000 rice varieties, 1,500 mango and banana varieties, and 4,500 varieties of brinjal. We bred our crops for diversity, nutrition, taste, quality, and climate resilience. Today we are growing a handful of chemically grown commodities, which are nutritionally empty, and full of toxins. We are facing a severe crisis of malnutrition with India emerging as 99th on the hunger index. 75% of the chronic diseases are "food style diseases."

In spite of the rich scientific and intellectual heritage based on food as health, India is rapidly emerging as the epicenter of chronic diseases including cancer, obesity and diabetes, and cardiovascular diseases—all largely related to food.

The malnutrition crisis and the food style disease epidemic is now a national emergency. It needs to be addressed urgently, by citizens, scientists and experts, and policy and decision makers. Health and disease begins in food, and food begins in agriculture and the soil.

There is an intimate connection between the biodiversity in the soil, the biodiversity of our plants, and the biodiversity in our gut, between ecological sustainability and health. Health is a continuum, from the Earth to our bodies. After all, we are made of the same *panch mahabhutas* as the Earth.

The soil, the gut, and our brain are one interconnected biome. Violence to one part triggers violence in the entire inter-related system. A healthy planet and healthy people are connected through nutrition, and nutrition-sensitive agriculture puts the nutrition cycle and diversity on our farms and diversity on our *thalis* at the center of agriculture.

When we apply urea to soil, the rich biodiversity of soil microorganisms that create the diversity of soil nutrients is destroyed, and the soil becomes diseased and desertified. Desertification is the death of soils.

In a similar way, when we eat poisons our gut mircobiome starts to get desertified. Because we are more bacteria than human, when the poisons we use in agriculture, such as pesticides and herbicides, reach our gut through food, they can kill beneficial bacteria and through it our health.

Nutrition-sensitive, biodiverse agriculture is at the heart of regenerating and rejuvenating the health of the planet and the health of people.

The IAASTD Report, the United Nations Millennium Ecosystem Assessment Synthesis Report, Tilman and other researchers have raised the issue of agricultural chemicals as significant contributors to global environmental change due to persistent and short-term environmental toxicants that are responsible for endocrine disruption, immune system diseases, and developmental toxicity (Aldridge 2003; Buznikov 2001; Cabello 2001; Cadbury 1997; Colborn 1996; Hayes 2002 & 2003; Qiao 2001; Short 1994; Storrs 2004; Tilman *et al.*, 2001; MA Report 2005; IAASTD 2008).

The damage caused by agricultural chemicals in the environment began to receive attention in the early 1960s when Rachel Carson wrote *Silent Spring* (Carson 1962). These chemicals were shown to persist and accumulate in the environment, causing mortality, birth defects, as well as mutations and diseases in both humans and animals.

In the 1990s the issue of chemicals adversely disrupting the reproduction of all species was made public through books like *Our Stolen Future* and *The Feminisation of Nature*. The peer-reviewed science that was summarized in these books showed that many chemicals, especially agricultural chemicals, were mimicking reproductive hormones like estrogen. These were causing serious declines in fertility by reducing the quantity and quality of sperm production and through damage to the genital systems, especially by feminizing male genitalia. (Colborn, Dumanoski & Myers 1996; Cadbury 1997).

The body of science showing that agricultural chemicals are responsible for declines in biodiversity and other environmental health problems continues to grow. These toxic chemicals now pervade the whole planet, polluting our water, soil, air and, most significantly, the tissues of most living organisms (Short 1994; Colborn, Dumanoski & Myers 1996; Cadbury 1997).

The public release of this information has had no effect in reducing the problem. Synthetic biocides (pesticides, fungicides and herbicides) have increased exponentially from when Rachel Carson wrote *Silent Spring* (Carson 1962). More than 7,200 registered biocide products are now used in Australian agriculture (Infopest 2004). This is significantly more than the nearly 1,400 pesticides have been registered by the Environmental Protection Agency (EPA) for agricultural and non-agricultural use in North America. (Reuben 2010)

Regulatory authorities do not acknowledge the multiple problems caused by widespread chemical use. The attitude by regulatory authorities and governments is that there are no environmental or health problems associated with current uses as the current regulatory regimes effectively manage their use.

In reality the regulation of chemicals in the environment is done in an ad hoc and inconsistent manner with widespread use of chemicals in some countries that are banned in others.

The issue that inadequate pesticide regulation is resulting in major environmental and human health problems has been validated by several recent studies. The most significant has been the Report by the US President's Cancer Panel. This report was written by eminent scientists and medical specialists in this field and it clearly stated that environmental toxins, including chemicals used in farming, are the main causes of cancers.

The report published by *The U. S. Department of Health and Human Services, The National Institutes of Health* and *The National Cancer Institute* raised many critical issues around chemical regulation. (Reuben 2010). These are:

+ **Most Chemicals have not been Tested for Safety**

  "Only a few hundred of the more than 80,000 chemicals in use in the United States have been tested for safety."

+ **Pesticides Linked to Many Types of Cancers**

  'Nearly 1,400 pesticides have been registered (i.e., approved) by the Environmental Protection Agency (EPA) for agricultural and non-agricultural use. Exposure to these chemicals has been linked to brain/central nervous system (CNS), breast, colon, lung, ovarian (female spouses), pancreatic, kidney, testicular, stomach cancers, and Hodgkin non-Hodgkin lymphoma, multiple myeloma, and soft tissue sarcoma. Pesticide-exposed farmers, pesticide applicators, crop duster pilots, and manufacturers also have been found to have elevated rates of prostate cancer, melanoma, other skin cancers, and cancer of the lip.

  'Approximately 40 chemicals classified by the International Agency for Research on Cancer (IARC) as known, probable, or possible human carcinogens, are used in EPA-registered pesticides now on the market.'

+ **Testing is Inadequate**

  Available evidence on the level of potential harm and increased cancer risk from many environmental exposures is insufficient or equivocal. The panel is particularly concerned that some known and suspected carcinogens' impact, mechanisms of action, and potential interactions are poorly defined.

'Meaningful measurement and assessment of the cancer risk associated with many environmental exposures are hampered by a lack of accurate measurement tools and methodologies. This is particularly true regarding cumulative exposure to specific established or possible carcinogens, gene-environment interactions, emerging technologies, and the effects of multiple agent exposures.'

+ **Current Testing Fails to Accurately Represent Human Exposure to Harmful Chemicals**

  'Some scientists maintain that current toxicity testing and exposure limit-setting methods fail to accurately represent the nature of human exposure to potentially harmful chemicals. Current toxicity testing relies heavily on animal studies that utilize doses substantially higher than those likely to be encountered by humans.'

+ **Testing Fails to Account for Harmful Effects of Low Doses**

'These data - and the exposure limits extrapolated from them - fail to take into account harmful effects that may occur only at very low doses. Further, chemicals typically are administered when laboratory animals are in their adolescence, a methodology that fails to assess the impact of in utero, childhood, and lifelong exposures.'

+ **Testing Fails to Account for Combinations of Chemicals'**

'In addition, agents are tested singly rather than in combination.'

'Single-agent toxicity testing and reliance on animal testing are inadequate to address the backlog of untested chemicals already in use and the plethora of new chemicals introduced every year.'

+ **Children are at Special Risk for Cancer Due to Environmental Contaminants**

'They [children] are at special risk due to their smaller body mass and rapid physical development, both of which magnify their vulnerability to known or suspected carcinogens, including radiation. Numerous environmental contaminants can cross the placental barrier; to a disturbing extent, babies are born "pre-polluted." Children also can be harmed by genetic or other damage resulting from environmental exposures sustained by the mother (and in some cases, the father). There is a critical lack of knowledge and appreciation of environmental threats to children's health and a severe shortage of researchers and clinicians trained in children's environmental health.'

+ **Children's Cancer Rates are Increasing**

'Yet over the same period (1975–2006), cancer incidence in US children under 20 years of age has increased.'

+ **Leukemia Rates Higher for Children Exposed to Pesticides**

'Leukemia rates are consistently elevated among children who grow up on farms, among children whose parents used pesticides in the home or garden, and among children of pesticide applicators. Because these chemicals often are applied as mixtures, it has been difficult to clearly distinguish cancer risks associated with individual agents.'

+ **Concern over Food Residues being too High**

'Only 23.1 percent of [food] samples had zero pesticide residues detected,

29.5 percent had one residue, and the remainder had two or more. The majority of residues detected were at levels far below EPA tolerances ... but the data on which the tolerances are based are heavily criticized by environmental health professionals and advocates as being inadequate and unduly influenced by industry.'

+ **Regulation of Environmental Contaminants is Inadequate to Protect from Harm**

'The prevailing regulatory approach in the United States is reactionary rather than precautionary. That is, instead of taking preventive action when uncertainty exists about the potential harm a chemical or other environmental contaminant may cause, a hazard must be incontrovertibly demonstrated before action to ameliorate it is initiated. Moreover, instead of requiring industry or other proponents of specific chemicals, devices, or activities to prove their safety, the public bears the burden of proving that a given environmental exposure is harmful.'

6.6 *Agricultural Chemicals in the Environment*

While the primary purpose of the report by the US President's Cancer Panel was to focus on human health, its findings have clear implications for the wider environment and all biota. These findings are consistent with a large and ever-growing body of peer-reviewed science showing the widespread and serious problems that are occurring in the environment from agricultural chemicals.

The Great Barrier Reef (GBR) is the proverbial canary in the coal mine on this issue. A report from the Great Barrier Reef Marine Park Authority stated:

> "Water quality in the Great Barrier Reef is principally affected by land-based activities in its adjacent catchments, including vegetation modification, grazing, agriculture, urban development, industrial development and aquaculture. Nutrients, sediments and pesticides are the pollutants of most concern for the health of the Great Barrier Reef." (Prange *et al.*, 2007)

This report and other studies showed that residues of several herbicides along with increases in the levels of nitrates and phosphates from synthetic fertilizers and soil loss in farming are damaging the Great Barrier Reef. The serious damage that these chemicals are doing to the reef, along with the pressures caused by climate change, can potentially destroy this world heritage-listed ecosystem. This has led to the Queensland Government regulating the main agricultural industries to lessen the number of chemicals affecting the GBR.

Pesticides pollute our drinking water and air. In 1999, Swiss research demonstrated that some of the rain falling on Europe contains such high levels of pesticides that it would be illegal to supply it as drinking water (Pearce 1999). Rain over Europe was laced with atrazine, alochlor, 2,4-D, and other common agricultural chemicals sprayed onto crops. A 1999 study of rainfall in Greece found one or more pesticides in 90% of 205 samples taken. Atrazine was measurable in 30% of the samples (Charizopoulos & Papadoupoulou-Mourikidou1999).

The situation is similar in the US: "The U.S. Geological Survey's (USGS's) recent national monitoring study found atrazine in rivers and streams, as well as groundwater, in all 36 of the river basins that the agency studied. It is also often found in air and rain; USGS found that atrazine was detected in rain at nearly every location tested. Atrazine in air or rain can travel long distances from application sites. In lakes and groundwater, atrazine and its breakdown products are persistent, and can persist for decades" (Cox 2001).

Atrazine interferes with the endocrine system (Hayes 2002 & 2003; Storrs et al., 2004).

It causes tumors of the mammary glands, uterus, and ovaries in animals (US EPA 2002). Studies suggest that it is one of a number of agricultural chemicals that cause cancer in humans (International Agency for Research on Cancer 1999; Mills 2002).

6.7 *Environmental Health*

In experiments conducted by Warren Porter and colleagues at the University of Wisconsin- Madison, mice were given drinking water with combinations of pesticides, herbicides, and nitrates at concentrations currently found in groundwater in the USA. They exhibited altered immune, endocrine, and nervous system functions (Porter 1999).

Porter showed that the influence of pesticide, herbicide, and fertilizer mixtures on the endocrine system might also cause changes in the immune system and affect fetal brain development. Of particular concern was thyroid disruption in animals. This has multiple consequences, including effects on brain development, sensitivity to stimuli, ability or motivation to learn, and an altered immune function.

A later experiment by Porter and colleagues found that very low levels of a mixture of the common herbicides 2,4-D, Mecoprop, Dicamba, and inert ingredients, caused a decrease in the number of embryos and lives births in mice at all doses tested. Very significantly, the data showed that low and very low doses caused these problems (Cavieres, Jaeger & Porter 2002).

6.8 *The Inadequacy of Toxicology Models*

The report by the US President's Cancer Council and many other peer-reviewed studies show that the current toxicology models used by our authorities are inadequate in determining the safety of many chemical compounds.

Significant numbers of studies show that compounds considered to have very little toxicity in parts per million (ppm) have a range of adverse effects in parts per billion (ppb). These compounds disrupt hormone systems at levels 1,000 times lower than previous research stated was safe. Agricultural chemicals have been shown to mimic hormones such as estrogen, blocking hormone receptors or stopping hormone activity. These chemicals have been implicated in lower sperm counts, increases in breast, uterine, ovarian, testicular, and prostate cancers, and deformities in the genital-urinary tracts in a wide range of species, including humans (Colborn, Dumanoski & Myers 1996; Cadbury 1997).

An example of this is Atrazine—one of the world's most commonly used herbicides. Two peer-reviewed studies conducted by Tyrone Hayes showed that levels 1,000 times lower than currently permitted in our food and in the environment caused severe reproductive deformities in frogs (Hayes 2002 & 2003).

Sara Storrs and Joseph Kiesecker of Pennsylvania State University confirmed Hayes' research. They exposed tadpoles of four frog species to Atrazine. *'Survival was significantly lower for all animals exposed to 3 ppb compared with either 30 or 100 ppb… These survival patterns highlight the importance of investigating the impacts of contaminants with realistic exposures and at various developmental stages'* (Storrs, 2004).

Dan Qiao and colleagues of the Department of Pharmacology and Cancer Biology, Duke University Medical Center, found that the developing fetus and the newborn are particularly vulnerable to amounts of pesticide far lower than currently permitted by most regulatory authorities around the world. Their study showed that the fetus and the newborn possess lower concentrations of the protective serum proteins than adults (Qiao et al., 2001). A major consequence is a developmental neurotoxicity, where the poison damages the developing nervous system (Aldridge 2003; Buznikov 2001; Cabello 2001).

The scientists stated: *'These results indicate that chlorpyrifos and other organophosphates such as diazinon have immediate, direct effects on neural cell replication… In light of the protective effect of serum proteins, the fact that the fetus and newborn possess lower concentrations of these proteins suggests that greater neurotoxic effects may occur at blood levels of chlorpyrifos that are nontoxic to adult.'* (Qiao et al., 2001)

6.9 *Epidemiology and Scientific Testing*

Most of the biocides used in farming are synthetic chemicals that have never existed in nature before. Scientists are continuing to find serious unintended consequences on the environment and human health. An abundance of published scientific research links commonly used pesticides such as Malathion, Diazinon, Chlorpyrifos, and other organophosphates as well as the carbamates, synthetic pyrethroids, and herbicides to disruptions to the hormone, nervous, and immune systems. They are also linked to cancers such as pancreatic, colon, lymphoma, leukemia, breast, uterine, and prostate. Autoimmune diseases linked include asthma, arthritis, and chronic fatigue syndrome.

This paper cannot detail them all; however, a few examples of common herbicides follow. A case-controlled study published in March 1999 by Swedish scientists Lennart Hardell and Mikael Eriksson showed that non-Hodgkin's lymphoma (NHL) is linked to exposure to a range of pesticides and herbicides (Hardell & Eriksson1999).

Hardell and Eriksson published an earlier study linking phenoxy herbicides such as 2,4-D to non-Hodgkin's lymphoma (NHL) in 1981. Before the 1940s, non-Hodgkin's lymphoma was one of the world's rarest cancers.

Now it is one of the most common. Between 1973 and 1991, the incidence of non-Hodgkin's lymphoma in the US increased at a rate of 3.3% per year, to become the third fastest-growing cancer. In Sweden, the incidence of NHL has increased at the rate of 3.6% per year in men and 2.9% per year in women since 1958 (Harras, 1996).

One of the biocides linked to NHL by the Hardell and Eriksson study is glyphosate. A previous study in 1998 had implicated glyphosate in hairy cell leukemia (Nordstrom et al., 1998). Animal studies have also shown that glyphosate can cause gene mutations and chromosomal aberrations (Cox 2004).

Research has shown that glyphosate can cause genetic damage, developmental disruption, morbidity, and mortality in amphibians at normal levels of use (Clements, Ralph & Petras 1997; Howe *et al.*, 2004; Lajmanovich, Sandoval & Peltzer 2003).

Clements and colleagues published a study showing damage to DNA in bullfrog tadpoles after exposure to glyphosate. The scientists concluded that its *"genotoxicity at relatively low concentrations"* was of concern (Clements, Ralph & Petras1997).

A 2003 study showed that a glyphosate herbicide caused both mortality and malformations in a common tadpole (Lajmanovich *et al.*, 2003). A 2004 study conducted by biologists at Trent University, Carleton University (USA), and the University of Victoria (Canada) showed that environmentally relevant concentrations of several glyphosate herbicides caused developmental problems in a common tadpole. The exposed tadpoles did not grow to the normal size, took longer than normal to develop, and between 10 and 25 percent had abnormal sex organs (Howe *et al.*, 2004).

Professor Don Huber's research shows that glyphosate causes major disruptions to the soil ecology. It kills or harms many species generally regarded as beneficial, such as cyanobacteria that have important roles in fixing nitrogen into the soil. It also stimulates other species, especially several major pathogens, which attack and damage other species. (Huber 2010).

7.1 *Farmer Livelihood as the Foundation of our Food Systems*

In today's agriculture, policies are driven by multinational corporations. Their profits come first, and this is why we see a rapid unfolding of social and ecological dilemmas, which threaten farmers and the very future of global crop cultivation.

The roots of the modern agriculture crisis lies in the Green Revolution—a paradigm based on chemicals that were used to fight wars, which manifested as a war against the Earth, farmers, and citizens. We now face a lack of diverse, nutritious crops in our food supply.

Environmental and social sustainability have been undermined by the dominant paradigm of industrial agriculture, which only measures growth in globalized trade. When development and economic growth are perceived exclusively in terms of capital accumulation, it often occurs through the destruction of sustainable local economies.

Farmers' livelihoods are also being undermined through an economically unsustainable and inequitable globalized agriculture model where corporate profits grow at the expense of farmers' livelihoods, the globalization of industrial agriculture leads to dependence on the chemicals and seeds, and international trade leads to increased dependence on growing a handful of commodities traded globally which have a high cost of production but low cost of sale.

Recognizing the need to promote sustainable agricultural systems and strengthen internal inputs, Agenda 21 of the United Nations Conference on Environment and Development states:

> "[I]ncreased use of external inputs and development of specialized production and farming systems tend to increase vulnerability to environmental stresses and market fluctuations. There is, therefore, a need to intensify agriculture by diversifying the production system for maximum efficiency in the utilization of local resources while minimizing environmental and economic risks."

Globalization has transformed agriculture into a market for expensive seeds and chemicals, while producing commodities at artificially low prices.

Continuing on the path shaped by global interests will only make this crisis worse.

Making the small farmer dispensable has been the intention of industrial agriculture. Less farmers means more automated systems that require a higher demand of fossil fuels and chemicals. The end of farmers is the end of food as we know it.

7.1 *Addressing Farmer Suicides and Honoring the Lives Lost*

More American farmers die to suicides than any other unnatural cause. As a group, farmers are likely to kill themselves at least three times as often as the general population. The epidemic of farmer suicides has now reached India. Hundreds of thousands of farmers have killed themselves because globalization of industrial agriculture has implied costly inputs and unpayable debts.

On June 4, 1984, farmers of Punjab were going to stop the supply of grain to Delhi because of their falling income. From June 1-10, 2018, Indian farmers went on strike and stopped the supply of vegetables and milk because of falling prices and declining incomes.

Falling incomes, rising production costs, indebtedness, and suicides have become the pattern of the socio-economic farming landscape in India and across the world in regions where industrial chemical farming is used.

A public trial was held on September 24–25, 2000 at Bangalore to understand the root causes of farmers suicides by listening to the victims' family members, offering solidarity, and making the problem visible.

Increasingly, farming communities are losing their family members, driven to death by the increased cost of seeds, increased debts, and crop failures. There have been several cases, in which farmers had to sell their land and even their kidneys to pay off their loans, houses, or tractors, which have been mortgaged. If there is failure to pay back the loans, they are at risk of being arrested.

On September 8, 2006, nine farmers' union teams of Punjab organized a public hearing on farmers' suicides. I was invited as a member of the citizens' jury. The Diwan Hall of Gurdwara Haaji Rattan was overflowing with a sea of people, all family members of suicide victims. The farmers' organizations had collected information on 2,860 suicides, and mobilized family members to give evidence at the public hearing.

Sukhbir Singh of Chak Sadoke, Block Jalalabad, District Ferozepur was 42 years old when ended his life on October 26, 2003 by jumping into a river because he was unable to pay a debt of $25,000 despite selling his seven acres of land. He left behind a widow with two children. Harjinder Singh, 21 years old, of Ratla Thark lost his seven acres to moneylenders and ended his life by consuming pesticides. Jeet Singh, 60 years old, of the same village burnt himself to death. Hardev Singh, 28 years old, of Urmmat Puria in Hoga drank pesticide on July 12, 2002 when he could not clear his loan of $9,300 even after selling eight acres. Avatar Singh, 26 years old, of Machika village died after consuming pesticide on March 18, 2006. Jagtar Singh, 48 years old, of Doda in Mukstar left behind a widow and daughter after drinking pesticide to end his life. He had sold two acres to partially pay a debt of $2,000. Raghubir Singh, 28 years old, mortgaged four acres, could not clear his loan, ended life on April 28, 2004, by consuming pesticide. His mother, widow, and two children are left to struggle. There are several cases of farmers' suicide and to show how farmers are paying corporate led globalization with their lives, Navdanya produced its report "Seeds of Suicide."

One by one, the women came to share their pain, their loss, their tragedy. The names and faces were different, but the pain was shared—the preventable tragedy of farmers' lives being lost. Suicide by drinking lethal pesticide took the lives of thousands. These are some of their names:

Gurjit's husband Budh Singh
Baljit Kaur's husband Thail Singh
Karamjit's husband Bhola Singh
Manjit Kaur's husband Sunder Singh
Gurmeet's husband Gudu Singh
Paramjit's husband Pritpal
Gurdayal Kaur's husband Jarnail Singh
Sukhpal's husband Gurcharan Singh
Jeet Kaur's husband Gurmeet Singh
Malkeet's husband Nishatar,
Tel Kaur's husband Nirpal
Sarabjit's husband Prem Singh
Jagat Kaur's husband Balbir

Surjeet Kaur's husband Dilwar Singh
Kulwinder Kaur's husband Sindoore Singh
Manjir Kaur's husband Chattar Singh
Amarjeet's husbank Pappi
Jasbir's husband Nirpesh Singh
Sukhdev Kaur's husband Birpal
Paramjeet's husband Pappi Singh
Sukhdev Kaur's husband Balwant Singh

Daljit Kaur's husband Sumukh
Harbans Kaur's son Gurmeet
Baldev Kaur's son Mewa Singh
Beant Kaur's husband Jailer Singh
Tej Kaur's husband Buttu Singh
Jasbir Kaur's son Jagga Singh
Tej Kaur's husband Mitti Singh
Jasbir Kaur's husband Kishan Singh
Charanjeet Kaur's husband Mahadev Singh

As I heard the unending stories from widows of how they had lost their dear ones, their land, and their hopes in the vicious cycle of debt, my mind went back to 1984 during the Green Revolution because of the violence of extremism and terrorism that had overtaken this prosperous and proud land of five rivers ("Punj" is five, "ab" is river). Those who survive suicide in Punjab face another threat to their lives: cancer. Pesticides are designed to kill, and from Punjab to Bhopal, they have killed thousands. The 1984 gas leak in Bhopal that resulted in thousands of deaths was from the pesticide plant of Union Carbide.

Agrichemicals and seed from agrichemical corporations have created an agrarian crisis, trapping farmers in debt, driving them to suicide, and leaving those who remain with terminal health conditions. The table below gives the suicides since 1995 when globalization was imposed through the World Trade Organization (WTO).

Table 1: Farmer Suicide from 1995-2015

Year	Yearly Total from All India Suicides
1995	10,720
1996	13,729
1997	13,622
1998	16,015
1999	16,082
2000	16,603
2001	16,415
2002	17,971
2003	17,164
2004	18,241
2005	17,131
2006	17,060
2007	16,632
2008	16,196
2009	17,368
2010	15,964
2011	14,027
2012	13,754
2013	11,772
2014*	5,660*
2015**	8,007**

* Total 1995-2014 = 3,02,126
** Total 1995-2015 = 3,10,133
* The actual figure for 2014 is 12,360, as NCRB did not include agricultural laborers and the actual figure for 1995-2014 is 3,08,126.
** The actual figure for 2015 is 12,602 as NCRB did not include agricultural laborers and the actual figure 1995-2015 is 3,20,728.

We need to make a transition from the suicide economy and false model of food security which is killing our children through malnutrition, our farmers through debt and suicides, and many others because of the unnecessary use of toxic poisons in farming. We can reduce global hunger, poverty, and disease by reducing our dependance on toxins.

There is a non-violent alternative to the Green Revolution—biodiverse and organic farming. Contrary to the claims by the chemical industry, as discussed previously, biodiverse ecological systems produce more food and nutrition than chemical monocultures. In this section, we will provide the instructions for achieving higher net incomes for farmers based on the experience and research of Navdanya International.

7.2 *The Impacts of Food Systems Globalization*

The policies of globalization and trade liberalization have created the farm crisis at three levels:

1. A shift from "food first" to "trade first" and "farmer first" to "corporation first" policies.
2. A shift from diversity and multi-functionality of agriculture to monocultures and standardization, chemical and capital intensification of production, and deregulation of the input sector, especially seeds, leading to rising costs of production.
3. Deregulation of markets and withdrawal of state from effective price regulation, leading to collapse in farm commodity prices.

Globalization is supposed to improve efficiency and productivity, and it is supposed to increase competition and choice. In the area of food and farming, globalization is based on three assumptions:

1. Globalization will lead to a spread of the most productive and efficient farming systems. It will therefore lead to increased food production as well as conserve land, water, and biodiversity.
2. Globalization will lead to increased access to the best food and decrease food scarcity and hunger. It will improve food security and safety.
3. Globalization will lead to increased competition and improve farmers' income and conditions by increasing their bargaining power and choice.

However, the reality of globalization and its impacts on food and farming is contrary to this mythology of prosperity and abundance.

Globalization of food and agriculture has led to:

1. Unsustainable industrial agriculture.
2. Rural poverty, and destruction of farmers' livelihoods and dramatic decline in farmers' incomes.
3. More hunger.

Monocultures

All around the world, food production is changing in a negative way. Farmers are spending more on inputs and getting lower prices for their produce. This is usually explained as being a result of surpluses and overproduction; however, the true problem is that the food in question is coming in the form of monocultures and monopolies.

One of the side effects of our economic system which promotes efficiency and profit is the push for a single, uniform crop. Modern agricultural technology has advanced this ability to plant one variety. The introduction of uniformity had been justified by the higher yield of a single crop to meet specific market demands.

Corporate giants determine the prices of seeds and chemicals, which can lower the price of produce in the short term. But in the long run, the control of these prices by monopolies eventually leads to high food prices.

Contrary to common perception, GMOs and chemical intensive farming systems lead farmers to spend more than they earn. Corporate, industrialized, globalized agriculture creates debt, creates reliance on costly inputs, and then purchases commodities below cost of production. It traps farmers in a system where they are paying more than they earn as their costs of production continuously rise, and the amount they earn for their products continuously falls. Even though US farm exports are booming, farmers cannot survive and often struggle to meet their own subsistence needs.

It is believed that industrial agriculture produces more food, and increased production leads to lower prices. When viewed in terms of total output of diverse crops as opposed to a single crop (defined as total nutrition), industrial agriculture does not outproduce farms using sustainable and biodynamic methods. Low prices are controlled by monopolies and are not accurately representative of total farm productivity. Wheat and rice contain the least nutritional value but make up most of the world's agriculture, displacing more nutritionally valuable crops. If measured by nutrition per acre, the Green Revolution didn't increase nutrition availability, especially for the marginal and poorest populations.

In a system of vertical integration from inputs to procurement, distribution, and processing, the same multinational corporations control:

A. The price at which inputs are sold to farmers.

B. The price at which commodities are bought from farmers.

C. The price at which the agriculture products are sold to consumers.

The income for farmers is (B) − (A). Since the costs of inputs are higher than the price of commodities, farmers' incomes are in the negative which leads to accruing debt. Simultaneously, the profit for corporations is (A) − (B). The more expensive the inputs sold to farmers, and the lower the prices they purchase from farmers, the higher their profit margin.

If governments did not give direct support to farmers, this system would collapse, and instead of selling inputs and buying commodities, corporations would have to bear the costs of production, and food prices would raise dramatically for them to keep making profit.

The industrial model of agriculture based on high costs of industrial inputs, the reduction of diverse economies, and the corporate economy of globalized trade have combined to create the rural distress we are witnessing worldwide.

Chemical industrial farming has created multiple economic problems:

+ Rising costs of production, causing debt
+ Falling net incomes for farmers because of globalization and dependence on a small amount of globally traded commodities
+ The true cost of food are hidden through subsidies and externalities.
+ Consumers are increasing offered unhealthy, chemically contaminated food at a price which does not cover the true cost born by nature and society

Low food prices, therefore, reflect the inefficient food system controlled by agribusiness monopolies who engage in price fixing and subsidies, threatening food sovereignty and undermining farmers' livelihoods.

Comparing Internal and External Inputs

Many Indigenous farming techniques are based on internal inputs—seeds come from the farm, soil fertility comes from environmentally beneficial compost and fertilizers, and pest control is built into crop designs through methods like companion planting.

Industrial agriculture replaces farm-level sustainability with dependance on external inputs (chemical fertilizers, pesticides, and herbicides) that have to be purchased at a high cost. The assessment of total farm productivity, or yield, does not consider the ecological effects of each additional input.

7.3 *Alternative Models of Economics and Sustainability*

Agriculture today has become highly unsustainable. We must look at the various dimensions of sustainability that consider human and ecological impacts:

A. **Natural resource sustainability**

Natural resource sustainability is based on the stability of agricultural ecosystems. This is the foundation of a nature-based economy which includes biodiversity, soil fertility, and water as the ecological capital for agriculture. Sustainability in nature involves the regeneration of nature's processes. Applying this principle in agricultural communities involves the regeneration and revitalization of the culture and local economy of production.

B. **Socioeconomic sustainability**

Socioeconomic sustainability can be measured through direct relationships between:

+ Society and the environment
+ The various individuals engaged in production
+ Producers and consumers

Socioeconomic sustainability measures the health of "the people's economy" or "the economy of sustenance," in which human needs are prioritized as human rights. The people's economy includes the diverse costs and benefits, both material and financial, that farming communities derive from agriculture.

Debt and farmer suicides are clear indicators of the unsustainability of the people's economy. The epidemic of farmer's suicides in India is concentrated in regions where chemical intensification has increased costs of production and cash crop monocultures face a decline in prices and incomes due to globalization. The high costs of production are the most significant reason for rural indebtedness. While farmers' incomes are falling, the price of food is increasing. How we produce and distribute our food is the single biggest determinant for the sustainability of nature's economy and people's economy, including farmers' livelihoods and citizens' health.

There are quite clearly two different meanings of sustainability. The real meaning refers to nature and societal sustainability. It involves a recovery of the recognition that nature supports our lives and livelihoods, it is the primary source of sustenance. Sustaining nature implies maintaining the integrity of nature's processes, cycles, and rhythms.

There is a second kind of "sustainability," which refers to market and corporate profits. It involves maintaining supplies of raw material for industrial production and long-distance global consumption. It involves sustainability of profits through an extractive economy. In this meaning, markets grow while the soils and rural communities are impoverished.

Growth of the Market Economy at the cost of People's Economy
and Nature's Economy

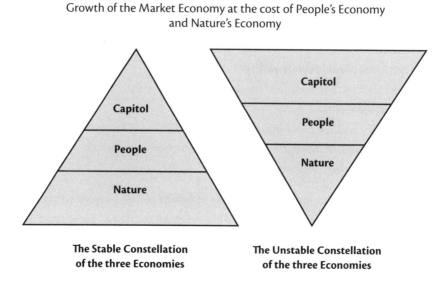

Figure 1: Biodiversity of Economies. Source: V. Shiva, Ecology and the Politics of Survival, SAGE Publications and UNU Press, 1991.

From Linear Extractive Economies to Circular Economies

One of our seminal contributions to fair trade practices has been the creation of fair trade food communities, connected through biodiversity, justice, and direct connection.

The diversity of rices, wheats, millets, cracked wheat, breakfast cereals, cookies etc., reflects our commitment to conserving local seeds and biodiversity, practicing water prudent agriculture, and ensuring the livelihoods of small farmers and women in the face of globalization.

Through ecological fair trade, we are combining sustainability and social justice. Through biodiversity, we are sowing the seeds of freedom, prosperity and peace, of living economies and living democracies. This is real fair trade—based on freedom, for nature and freedom for people.

The dominant model of the economy no longer has its roots in ecology, but exists outside and above ecology, disrupting the ecological systems and processes that support life. The unchecked conquest of resources is pushing species to extinction and has led ecosystems to collapse, while causing irreversible climate disasters.

Similarly, economy, which is part of society, has been placed outside and above society, beyond democratic control. Ethical values, cultural values, spiritual values, and values of care and co-operation have all been sidelined by the extractive logic of the global market that seeks only profit. Competition leaves no room for cooperation. All values that arise from our interdependent, diverse, and complex reality have been displaced or destroyed. When reality is replaced by abstract constructions created by the dominant powers in society, manipulation of nature and society for profits and power becomes easy. The welfare of real people and real societies is replaced with the welfare of corporations. The real production of the economies of nature and society is replaced by the abstract construction of capital, and now digital finance. The real, the concrete, the life-giving gives way to the artificially constructed currencies with a uni-directional flow.

The linear extractive economy is based on extraction, commodification, and profits. It has no place for the care of nature and community. It leaves nature and society impoverished, be it through the extraction of minerals, extraction of knowledge through biopiracy, or extraction of 'genes' through genetic mining, extraction of data through 'data mining', or extraction of rents and royalties for seed, water, communication, or privatized education and health care. It creates poverty, debt, and displacement. It creates waste—waste as pollution, wasted resources, wasted people, wasted lives.

Real wealth is our capacity to create, produce, and make what we and our communities need to ensure our well-being. Well-being is the original meaning of wealth, not money. Work creates wealth. As co-creators and co-producers with nature we protect the Earth's wealth creating capacities and enhance our own. We create real wealth when we live as Earth Citizens.

Life and its vitality in nature and society is based on cycles of renewal and regeneration of mutuality, respect, and human solidarity. The relationship between soil and society, between nature and culture, is a relationship based on reciprocity, on the Law of Return, of giving back.

The ecological Law of Return maintains the cycles of nutrients and water, and hence the basis of sustainability. For society, the Law of Return is the basis of ensuring justice, equality, democracy, and peace.

Regenerative, renewable, sustainable economies that enhance nature's well-being and ours are based on the Law of Return—of giving back in gratitude and deep awareness that we are the web of life and must take care of it.

They are therefore circular economies that are aware of, and maintain nature's cycles. All ecological crises are the rupture of nature's cycles, and the transgression of what have been called "planetary boundaries." When we give back organic matter to nature, she continues to give us food. The work in giving back is our work. Giving us food is nature's complex work—through her soil, her biodiversity, her water, the sun, and the air.

In a circular economy we give back to nature and society. We build nature's economy and the people's economy. Wealth is shared. Wealth is distributed. Wealth circulates.

In real economies, plants grow, soil organisms grow, children grow in well-being and happiness.

The circular economy replenishes nature and society. It creates enoughness and well being for all. In the care of the Earth and society, diversity of meaningful and creative work is possible. It is based on nature's Law of Return. In nature, there is no waste, no pollution.

When economies are circular, every being and every place, is the center of the economy, and nature and society evolve and emerge from multiple self-organized systems, like the trillions of cells in our body.

Circular economies as living economies are by their very nature biodiverse, spanning from the intimate and local, to the global and planetary.

Through biodiverse ecological food systems and circular economies, we can end farmer suicides and reverse agrarian distress. Navdanya farmers have increased their incomes tenfold through seed sovereignty, food sovereignty, and cooperation among themselves, as well as with those who eat the diverse, healthy food they produce.

The circular logic of Law of Return, mutuality, reciprocity, and regeneration just as self-organized systems evolve in and through diversity, self-organized economies are diverse. For 30 years we have connected seed to table and farmers to eaters, creating circular cooperative economies as alternatives to the linear extractive economies through which corporations exploit the soil, the farmers by selling poisons, and citizens by selling degraded and fake food.

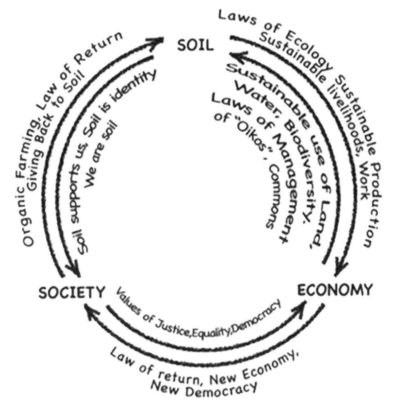

Figure 2: The circular logic of Law of Return, Mutuality, Reciprocity and Regeneration

Table 2: Average production and total amount of Baranaja, Navdanta, Septarashi and Punchranga v/s Monocropping growing at Navdanya Farm

S.No.	Name of the Crops	Average production/ha.(Kg.)	Average Rate/Kg	Total Amount Rs.
YEAR: 2004-2005				
BARANAJA				
1.	Bazara	440.00	8.00	3520.00
2.	Maize	1280.00	8.00	10240.00
3.	Sefed Chemi	600.00	25.00	15000.00
4.	Aongal	360.00	20.00	7200.00
5.	Mandua	600.00	10.00	6000.00
6.	Jhangora	440.00	15.00	6600.00
7.	Urd	600.00	20.00	12000.00
8.	Navrangi	680.00	20.00	13600.00
9.	Koni No.1	280.00	10.00	2800.00
10.	Lobia	600.00	20.00	12000.00
11.	Till	400.00	30.00	12000.00
12.	Koni No.2	340.00	10.00	3400.00
Total		6620.00		1,04360.00
MONOCULTURE				
1.	Maize	5400.00	8.00	43,200.00
NAVDANYA				
1.	Till	400.00	30.00	12000.00
2.	Sefed chemi	720.00	25.00	18000.00
3.	Mandua	1120.00	10.00	11200.00
4.	Dholiyia dal	640.00	20.00	12800.00
5.	Sefed Bhatt	760.00	15.00	11400.00
6.	Lobia	800.00	20.00	16000.00

Continued on next page.

S.No.	Name of the Crops	Average production/ha.(Kg.)	Average Rate/Kg	Total Amount Rs.
		YEAR: 2004-2005		
7.	Jhongora	520.00	15.00	7800.00
8.	Maize	560.00	8.00	4480.00
9.	Gheat	480.00	25.00	12000.00
Total		6000.00		1,05680.00
MONOCULTURE				
1.	Mandua	3600.00	10.00	36000.00
SEPTRASHI				
1.	Urd	600.00	20.00	12000.00
2.	Moong	520.00	25.00	13000.00
3.	Mandua	560.00	10.00	5600.00
4.	Sefed Bhatt	680.00	15.00	10200.00
5.	Dohyalya Dal	560.00	20.00	11200.00
6.	Maize	680.00	8.00	5440.00
7.	Lobia Dal	600.00	20.00	12000.00
Total		4200.00		69440.00
MONOCULTURE				
1	Urd	2400.00	20.00	48000.00

Table 3: Kharif / summer crop production in one acre

S. No.	Name	Production /kg	Personal Consumption	For Market	Value Saved	Market Rate
1.	Wheat	350	350		6300	16
2.	Mustard	50	50	-	2250	45
3.	Potato	300	100		3000	30
4.	Pea	50	20		1000	50
5.	Barley	40	40		720	18
6.	Raddish	200	20	180	500	25
7.	Rai	50	20	30	500	25
8.	Fenugreek	100	5	95	175	35
9.	Carrot	80	20	60	900	45
10.	Turnip	80	10	70	200	20
11.	Spinach	50	10	40	350	35
12.	Broadbean	50	10	40	400	40
13.	Oat (Fodder)	300	300		6000	20
14.	Barsim (Fodder)	400	400		8000	20
I5.	Linseed	8	1	7	100	300
16.	Cauliflower	45	20	25	800	40
17.	Knol Khol	30	10	20	250	25
18.	Camomile	2		2	0	800
19.	Stevia	0.5		0.5	0	500
20.	Onion	100	50	50	2000	40
21.	Garlic	10	5	5	800	160
22.	Coriander	20	5	15	250	50
	Total				34495	
					$ 515	

Total Rs/-	Seed Qt. Kg	Personal Use	Value Saved	For Market	Market Rate	Total Rs.
0	0	0	0			
0	0	0	0			
0		0	0			
0	5	5	600		120	600
0	0	0	0			
4500	3	5	1000		200	1000
750	0.5	0.5	100		200	100
1325	8	2	90	6	45	360
2700	2	0.5	175	2.5	390	700
1400	1	0.5	100	0.5	200	200
1400	3	0.5	75	2.5	150	450
1600	5	1	30	4	80	400
0	10	10	180		18	180
0			0			
200			0			
1000			0			
500			0			
1600			0			
250			0			
2000			0			
800			0			
750	3	1	200	2	200	400
23275				2600		4390
$ 348				$ 39		$ 66

Table 3: Kharif / summer crop production in one acre (continued)

S. No.	Name	Total Production /kg	Personal Consumption	For Market	Value Saved	Market Rate
1.	Til	7	2	5	160	80
2.	MakkaChari (Fodder)	2000	2000	-	40000	20
3.	Nav Rang	20	20		2000	100
4.	Paddy	250	250		6250	25
5.	Malle	150	50	100	1000	20
6.	Bhindl	70	20	50	600	30
7.	Fox tail Millet	5	5		200	40
8.	Banyard Millet	2	2		40	20
9.	Dhecha	2	2		300	150
10.	Turmeric	500	100	400	8000	80
11.	Ginger	100	5	95	375	75
12.	Arbi	100	20	80	800	40
13.	Lobia	5	5		200	40
14.	Bajra	10	10		200	20
IS.	Khira	50	20	30	500	25
16.	Ragi	80	80		2400	30
17.	Brown Bhatt	5	5		500	100
18.	Sponge Gourd	80	20	60	700	35
19.	Bottle Gourd	70	25	45	750	30
20.	Sweet Potato	20	2	18	110	55
21.	Ararote	10			0	40
22.	Bhanjiir	3	1	2	200	200
23.	Discoria	10	3	7	150	50
24.	Chillies	0.5	0.5		75	150
25.	Brinjal	10	10		400	40
26.	Herbs	10	2	8	700	350
	Total				66610	

Total Rs/-	Seed Qt. Kg	Personal Use	For Market	Value Saved	Market Rate	Total Rs.
400	0	0	0	0	0	0
	6	2	4	240	40	240
0	0	0	0	0	0	0
0	0	0	0	0	0	0
2000	8	3	5	160	20	160
1500	10	2	8	4000	400	4000
0	0	0	0	0	0	0
0	0	0	0	0	0	0
0	0	0	0	0	0	0
32000	0	0	0	0	80	0
7125	0	0	0	0	50	0
3200	0	0	0	0	50	
	12			1440	120	1440
0	0	0	0	0	0	0
750	2	1	1	1000	500	1000
0	0	0	0	0	0	0
0	0	0	0	0	0	0
2100	5	1	4	750	150	750
1350	2	1	i	300	150	300
990	0	0	0	0	0	0
0	0	0	0	0	0	0
400	0	0	0	0		
350				0		
0				0		
0				0		
2800				0		
54965				7890		7890

On one acre we sustain 47 trees of 20 different species, cultivate 26 summer crops and 22 winter crops which can sustain a family of 5 and produce an income of Rs 1,61,200.

Table 4: Tree diversity

S.No	Name of the Trees	Number of Trees
1.	Bamboo	95 Shoot
2.	Kamla	3
3.	Mulberry	12
4.	Jamun	3
S.	Kachanar	5
6.	Bhimal	2
7.	Tun	2
8.	Bakayan	6
9.	Retha	1
10.	Gulmohar	2
11.	Anjeer	1
12.	Saijan	1
13.	Kapoor	1
14.	Arokera	1
I5.	Sagwan	2
16.	Banana	6
17.	Papaya	1
18.	Sesam	1
19.	Kadam	1
20.	Kern	1

S. No	Name	Production qt/Kg	Personal Consumption	Value Saved	For Sale	Market Rate	Total Rs.
1.	Jamun	50	10	250	40	25	1000
2.	Mulberry	40	5	100	35	20	700
3.	Banana	5 dz	5dz			50/dz	
4.	Papaya	60	5	200	55	40	2200
Total				550			3900 ($58)

For three decades, Navdanya has been working to evolve paradigms and practices for regenerating farmers' livelihoods and increase their incomes while also regenerating the earth and growing healthy and nutritious foods.

Navdanya farmers practice biodiversity-based agroecology and therefore grow food without purchasing external inputs. They practice what is often also called "zero budget" farming. Being sovereign in seed, and in biodiversity-based organic renewal of soil fertility and pest management, they have zero expenditure in purchasing external chemical inputs. Being ecological, they are debt free.

All Navdanya farms are biodiverse farms which contribute to ecological functions that are alternatives to chemicals. They also provide biodiverse outputs. Farmers are thus never in the distressing situation that farmers growing chemical monoculture commodities are in since regenerative, biodiversity-based, organic agriculture produces high-quality and diverse foods.

The Navdanya philosophy is *Bija Swaraj* and *Anna swaraj* through *Jaivik Kheti*—regenerative organic farming. Navdanya farmers are therefore also sovereign in their pricing, distribution, and marketing systems through biodiversity of economies.

Consequently, their incomes are 10 times higher than farmers that are forced to grow commodities in chemical monocultures.

The Navdanya model for ecological/organic farming focuses on biodiversity. Navdanya means nine seeds as well as new gift. The most significant contribution by Navdanya has been the promotion of biodiversity-based productivity for small farmers, which combines ecological conservation with economic production.

At a time when GMO seeds are being offered as a miracle, just as the HYV seeds were introduced as a miracle, during the Green Revolution, Navdanya has conserved the open-pollinated farmer's varieties, reintroduced them in production systems, and enhanced both productivity and rural incomes.

The industrial, corporate globalized model of agriculture is based on the production of a few globally traded commodities with high external inputs. Thus farmers growing Bt cotton spend more than they earn and are pushed into debt, and in extreme cases to suicide.

On the other hand, biodiversity intensification with native seeds increases both output and incomes.

Monocultures produce more corporate control on production and distribution of commodities, not more food. They facilitate corporate control over agriculture by making farmers dependent on monopoly markets and high cost inputs. They create profits for corporations, which sell costly inputs and buy cheap commodities through contract farming. For farmers, they translate into a negative economy of high costs and low returns, which leads to debt, suicides and landlessness.

A study conducted by Navdanya in four districts of West Bengal shows that biodiversity in the same soil and climatic regimes proves economically more efficient than chemical intensive farming systems. The study shows that the net value of the annual production of an average biodiverse farm is uniformly more than that of an average monoculture farm. The selected biodiverse farms of East Medinipur district are sown to a wide range of crop diversity, both under sequential rotation and intercropping. Some of these farms—mostly smaller than a hectare in size—grow over 50 types of crops excluding rice. Rainfed farms of Bankura district are comparatively less diverse, hardly exceeding 14 crops a year including rice. The irrigated monoculture farms, by contrast, grow 2 rice varieties in Bankura district and three rice varieties (all HYV) in East Medinipur district. Monoculture farms of East Medinipur appear to be less productive in spite of three rice crops than those of Bankura with two rice crops. Farmers explain this to reflect the "farm fatigue" from monoculture and intensive use of agrochemicals—an essential feature of industrial agriculture.

A remarkable finding was that the relative value of the farm produce seems to increase significantly with greater diversity of crops. This unimodal distribution of the value of net farm profit (difference between the output and input value) per unit area *vis-a-vis* crop diversity becomes clear when the net profit and crop species numbers are both natural log-transformed. The regression slope is 0.5893, which is significant at 99.9% level of confidence.

The data contradicts the prevailing mainstream agronomic conjecture that intensive cropping of a monoculture crop would enhance productivity of the land. A majority of farmers in Bankura and Medinipur have now realized that over years, the yield of the monoculture farms is unsustainable. Many of these farmers have reverted back to traditional farming systems involving folk crop varieties.

Some of them have experimented with a hybrid system of rotational crop-ping of a large number of "secondary" crops and a HYV rice. However, most of these mixed crop farmers reported that "the cost of inputs eat away the extra production of HYV rice", and that the best means to cut down on the extraneous inputs is to "give the land a recess" by growing vegetables and fruits for a few years before replanting it with rice (RFSTE, *Industrial vs Ecological Agriculture*, Deb 2004).

Thus, conservation of native seeds and biodiverse ecological farming has led to incomes which are 2-3 times higher than monoculture, and 8-9 times higher than industrial systems using genetically engineered seeds.

Table 5: Cost of production & net profit of Balbeer Singh, a Navdanya farmer

Year	Urea/Bigha	DAP/Bigha	Potash/Bigha	Cow Dung Manura/Bigha
1994 - 1995	10 kg (100%)	10 kg (100%)	2 kg (100%)	2 qt (20%)
1995 - 1996	8 kg (80%)	8 kg (80%)	20%	3 qt 30%)
1996 - 1997	4 kg (40%)	4 kg (40%)	Nil	20 qt (100%)
1997 - 1998	Nil	Nil	Nil	40 qt (200%)
1998 - 1999	Nil	Nil	Nil	20 Qt (100%)

(Source: Bolbeer Singh. Village Utireha, and Navdanya Records)

Continued on next page.

Table 6: Cost of production & net profit of Balbeer Singh, a Navdanya farmer

Year	Wheat Yield/Bigha	Cost of Agrochemical	RiceYlald / Bigha
1994 - 1995	1.60 qt.	100	1.8
1995 - 1996	1.08	68	0.90
1996 - 1997	0.98	32	0.92
1997 - 1998	1.8	Nil	2.00
1996 - 1999	2.2	Nil	2.50
2004* - 2005	2.5	Nil	3.0

(Source: Bolbeer Singh. Village Utireha, and Novdanya Records)

The above table shows that a farmer in his 0.5 bigha of land (12.5 bigha = 1 ha) by doing multi cropping was able to earn a net profit of Rs. 3060. Cost of production was estimated to be Rs. 1200 for one year, which includes the man-days of the farmers as well as FYM from his own farm, although he did not spend any money for cultivation.

If we calculate the net income for one hectare, the farmer was able to make as much as Rs. 90,000, which is quite high. It is not easy to earn this much profit with any type of farming.

Thus it was observed that the more the diversity, so will be the income and profit of the farmer. This is contrary to what supporters of conventional farming tell the farmers in order to promote monocultures.

Another example is that of Yogambar Singh of Pulinda, a 65 year-old farmer who tells his story with great interest: "*I have about 40-nali of land and I am solely dependent on agriculture for my livelihood. I have no other source of income.*" He shared that he had been using extensive chemicals sometime before 1995. After joining Navdanya, he stopped using chemicals on his land and has been practicing organic farming for over 10 years. He goes on to say: "*I am an illiterate person, but I know farming. I used chemicals for a few years in my fields, which really deteriorated the soil fertility as well as texture and quality of the soil.*" In his life, he has married three times, and reports that he was only able to marry because of the income of his farm. He was also able to purchase 2 taxis for his son. His annual net income is at least Rs 70,000 excluding expenditures, most of which is for his labor, FYM, or his home-made composts.

Comparative analysis of his two fields, one irrigated and another non-irrigated, this was the primary example Yogambar Singh shared: that from non-irrigated fields, a farmer could earn equal or even more than that of an irrigated field. According to him, he has been convinced that only hard work and organic farming practices could earn high returns, not the intensive use of agrochemicals.

🌿 FIELD 1 IRRIGATED

Cost of production – Rs 1000.00

Gross Income – Rs 2745.00

Less Expenditure – Rs 1000.00

Net Income – Rs 1745.00
per (0.75 Nali)
Or Rs. 69500.00 per Ha

🌿 FIELD 2 UN-IRRIGATED

Cost of Production – Rs 600.00

Gross Income – Rs 2123.00

Less Expenditure – Rs 600.00

Net Income – Rs 1523.00 (0.5 Nali)
Or Rs. 951887.50 per Ha

In 1995, Navdanya evolved the metric of biodiversity-based productivity in place of yields of chemical monocultures commodities. 200 biodiverse farms were studied in four regions in four diverse agroecosystems—the mountain ecosysems of Uttarakhand and Sikkim; the desert ecosystem of Rajasthan; and the Western Ghats of Kerala.

The biodiversity output was used to estimate the total economic value on small farms according to *mandi* prices and retail prices. The *mandi* prices and the retail prices of diverse foods were collected by Navdanya in the month of February 2013 for calculating the wealth per acre on biodiverse farms compared with chemical monoculture farms. We take these rates as reference rates. The farm output, which was measured in terms of biodiversity, was converted to monetary terms, if the farmer sold all his produce. The final summary of results was as follows:

By taking market retail rates we assessed that the additional income generated by bioidverse organic farming was Rs 66,197 compared to Rs 33,037 for chemical monocultures per acre farmland.

The increase in income through biodiversity is therefore 66197–33037= Rs 33,160/acre

Bioidversity has thus ensured doubling of farm incomes.

304 Agroecology and Regenerative Agriculture

In India we have 45,22,02,848 acres of cultivable land. By directing all of it to biodiverse organic farming, the amount of extra incomes farmers generate for themselves would be: 45,22,02,848 x 33,160 = Rs 1,49,95,04,64,39,680 = US$ 2,76,20,27,34,199 = US$ 276.2 Billion/Year (approx.) = 15% of Indian GDP in the year 2011-12. The extra revenue generated may be equal to one sixth of the Indian GDP. This is the path to remove both poverty and hunger.

Biodiversity of Economies and Farmers' Market Sovereignty

For Navdanya, food is not a commodity but a commons. A farmer is a member of a food community as are those related to her/him through eating. Industrial agriculture and globalization has reduced food to a nutritionally empty toxic commodity, and has separated the modern consumer from the Eater and the realities of food cultivation. With long distance supply chains and giant corporations acting as the middlemen, farmer's crises are increased, and their livelihoods are driven below survival level.

Through biodiverse organic food, Navdanya has reclaimed food as both nourishment and heritage.

Navdanya has created alternative markets based on direct marketing of biodiverse products. This is distinctive as a fair trade movement because:

A. it creates markets through fair trade to protect biodiversity not just sell a commodity

B. it has created local, domestic markets in the South

C. it directly connects its producer members and consumer (co-producer members)

The direct marketing/fair trade initiatives of Navdanya are an alternative to the unfair trade driven by agribusiness and enforced by WTO and the World Bank. Corporate, industrialized, globalized agriculture creates debt, and uproots small farmers by exploiting them first through selling them costly chemicals and non-renewable seeds and then through purchase of commodities below cost of production. Globalized agriculture also uses uniformity and standardization to marginalize small farmers and biodiversity.

From monoculture to diversity

Navdanya, as a movement for conservation of biodiversity, uses fair trade to create diverse markets for diversity. The global market focuses on wheat, rice, corn and soya. The GMO dominated market focuses on corn, soya, canola, and cotton.

Navdanya has popularized "forgotten foods"—millets and psuedo cereal—which are resource prudent yet highly nutritional. Their population means less water is used, and more nutrition is produced. India could produce 400 times more food using the same land and water if priority was given to millets like Ragi (Finger Millet) and Jhangora (Barnyard Millet).

When the edible oil market was being reduced to soya and farmers growing oilseeds were losing their markets, Navdanya defended and promoted Indigenous oil seeds and their cold-pressing on Indigenous oil mills. More than nine varieties of mustard were conserved in addition to sesame, linseed, and niger. Fair trade in organic oilseeds and in cold-pressed edible oil has conserved biodiversity, and protected the livelihood option of farmers in arid areas like Rajasthan, both in oilseed production and edible oil extraction.

Navdanya has helped conserve more than 3,000 rice varieties, and it has created fair trade organic markets for unpolished brown rice, red rice, nine varieties of basmati, and nine varieties of aromatic rices.

Navdanya has also provided an alternative to the unhealthy monoculture of Coca Cola and Pepsi with its unique range of fruit juices—bel, ginger, rhododendron, seabuckthorn, malta and mint.

From centralization/concentration to decentralization and circular economies

Unfair trade imposes concentration in production, both by focusing on single commodities, as in India's Agriculture Export Zones, and by promoting concentration of landholdings by dispossessing small farmers.

Navdanya builds on the strength of biodiversity and small farmers. By conserving and promoting the uses of Indigenous biodiversity where the soils and agro climate best suit it, and by transforming biodiversity into the farmers' most important capital, Navdanya has evolved a horizontally organized network of decentralized, diversified, producer communities, who are simultaneously conservers of water and biodiversity. Navdanya builds on the Gandhian philosophy of ever expanding, never ascending circles. Gandhi's economic vision is best captured in his economic constitution of India.

"According to me the economic constitution of India and for the matter of that of the world, should be such that no one under it should suffer from want of food and clothing. In other words everybody should be able to get sufficient work to enable him to make the two ends meet. And this ideal can be universally realized only if the means of production of the elementary necessaries of life remain in the control of the masses. These should be freely available to all as God's air and water are or ought to be; they should not be made a vehicle of traffic for the exploitation of others. Their monopolization by any country, nation or group of persons would be unjust. The neglect of this simple principle is the cause of the destitution that we witness today not only in this unhappy land but in other parts of the world too."

From agribusiness-led to nature and people-led production and distribution

Biodiversity plus organic plus market sovereignty increases farmer's income by reducing input cost and middleman/corporate exploitation to zero.

Navdanya members are earning up to 10 times more when they get off the chemical and commodity treadmill, which traps them in rising cost of production and falling process for their produce.

Chemical Commodity Monocultures vs Biodiversity-Based Organic Farming and Fair Trade

Table 7: Net Incomes of monoculture vs ecological alternative (Biodiversity Based Organic Farming)

MONOCULTURES		ECOLOGICAL ALTERNATIVES	
Crop	Net Income (Rs)/ha	Crop	Net Income (Rs)/ha
Hybrid Rice	71862	Dehraduni Basmati	113032
Hybrid Corn	30659	Finger Millet	128150
Hybrid Soybean	2863	Rajma	267399
Green Peas	94715	Amaranth	367000
Bt Cotton	8403	Desi Kappas	23737
Average Net Income	41700	Average Net Income	179864

Redefining Productivity

The two dominant myths that promote industrial agriculture are that:

1. Monocultures produce more than biodiversity
2. Small farms are not productive or viable

Studies and our practice show both myths to be false.

On Navdanya's organic farm, we grow 12 crops, 9 crops, 7 crops, and 5 crops in mixtures. Biodiverse systems produce more food and higher incomes than monocultures. (See Table 7) A one acre model on the Navdanya farm shows that a small farmer growing biodiversity can produce both abundance and prosperity.

One of the major problems with the way that productivity is measured is that it doesn't take environmental costs or benefits into account. Economic calculations of agricultural productivity usually take into account only the yield of a particular crop per unit of land.

Productivity only looks at outputs and not how they are achieved. It also does not fully reflect additional financial costs when agricultural practices shift from diversity-based systems to monocultures.

This becomes problematic in farming systems, such as polycultures, where plants for various purposes (fuel or fodder) may be grown alongside the primary commodity crop.

Take one hectare of land, for instance, where there is a harvest of 4 tons of rice and 2 tons of straw. Since calculating productivity would apply only to the yield of the rice, a polyculture field could be defined as less productive than a monoculture field.

Sustainable cropping systems include a symbiotic relationship between soil, water, farm, animals, and plants. Ecological agriculture links them together in sustainable ways, where each individual component acts in interdependence with the rest, strengthening the whole system.

The idea that we can produce more food through industrial breeding and genetic engineering does not support traditional methods of farming, take farmers' livelihoods into consideration, or strengthen food sovereignty for the following reasons:

1. It focuses on partial aspects of single crops rather than the total yields of multiple crops in integrated systems
2. It calculates quantity per acre rather than nutrition per acre

3. It uses natural resources intensively and does not factor the cost of environmental impacts into the price of food.

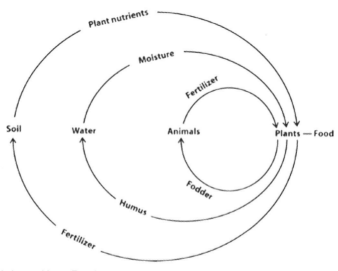

Figure 3: Internal Input Farming
Source: V. Shiva, *Violence of the Green Revolution*

Navdanya International's research on 22 rice-growing systems has shown that Indigenous systems are more efficient in terms of yields, labor use, and energy use (Table 8). This evaluation demonstrates that these methods result in long-term scarcity because the monoculture paradigm fails to take diversity of farming techniques and outputs into consideration.

Table 8: Comparing the inputs and outputs of pre-industrial, semi-industrial, and full industrial systems of rice cultivation (per hectare in one year)

LOCATION	FOSSIL FUEL INPUTS (%)	LABOR (PER CROP)	TOTAL LABOR (%)	TOTAL INPUT (GJ)	TOTAL OUTPUT (GJ)
Pre-industrial					
a) Dayak, Sarawak (1951)	2%	208	44%	0.30	2.4
b) Dayak, Sarawak (1951)	2%	271	51%	0.63	5.7
c) Kilombero, Tanzania (1957)	2%	170	39%	0.42	3.8
d) Kilombero, Tanzania (1967)	3%	144	35%	1.44	9.9
e) Iban, Sarawak (1951)	3%	148	36%	0.27	3.1

Continued on next page.

LOCATION	FOSSIL FUEL INPUTS (%)	LABOR (PER CROP)	TOTAL LABOR (%)	TOTAL INPUT (GJ)	TOTAL OUTPUT (GJ)
f) Lust'un Yunnan (1938)	3%	882	70%	8.04	166.9
g) Yits'un Yunnan (1938)	2%	1293	78%	10.66	163.3
h) Yuts'un Yunnan (1938)	4%	426	53%	5.12	149.3
Semi-industrial					
i) Mandya, Karnataka (1955)	23%	309	46%	3.33	23.8
j) Mandya, Karnataka (1975)	74%	317	16%	16.73	80.0
k) Philippines (1972)	86%	102	5.3%	12.37	39.9
l) Philippines (1972)	89%	102	4.1%	16.01	51.6
m) Japan (1963)	90%	216	5.2%	30.04	73.7
n) Hongkong (1971)	83%	566	12%	31.27	64.8
o) Philippines (1965)	98%	72	13%	3.61	25.0
p) Philippines (1979)	33%	92	16%	5.48	52.9
q) Philippines (1979)	80%	84	11%	6.90	52.9
Full-industrial					
r) Surinam (1972)	95%	12.6	0.2%	45.9	83.7
s) USA (1974)	95%	3.8	0.02%	70.2	88.2
t) Sacramento Calif (1977)	9%	3.0	0.04%	45.9	80.5
u) Grand prairie Ark. (1977)	95%	3.7	0.04%	52.5	58.6
v) Southwest Louisiana (1977)	95%	3.1	0.04%	48.0	50.8
w) Mississippi (Delta 1977)	95%	3.9	0.05%	53.8	55.4
x) Texas Gulf Coast (1977)	95%	3.1	0.04%	55.1	74.7

Industrial breeding and agricultural biotechnology are also responsible for reducing the nutritional value of our food. Nutrient-dense crops are replaced with high response varieties of lower nutritional value crops. The reduced calorie intake in our food is contributing to global food insecurity and the starvation of two-thirds of the world's population.

There are a number of strategies, which allow this inversion to take place and an illusion of growth to be created. Firstly, a monoculture paradigm looks only at one element of a system and treats an increase in the part as an increase in the whole system. Thus, focusing on yield alone increases of grain of individual cereals like rice or wheat, while reducing straw availability for fodder. It also reduces the nutritional content and quality in legumes, oil seeds, and greens.

A second strategy is to exclude the higher inputs from the resource equation and only focus on the single commodity output. Thus, the resource waste is not considered, and low resource use productivity is converted into high commodity productivity.

To assess the real productivity of a farming system from the farmer's perspective and the soil's perspective, we need to measure the biodiversity-based productivity and not just the price or yield of single commodity or single output. We also need to calculate:

A. the value of diverse outputs from diverse species and their diverse functions
B. the value of internal inputs provided by diverse farm outputs (e.g., straw for organic manure)
C. the costs of purchased inputs such as fertilizers, pesticides, herbicides
D. the ecological costs of external chemical inputs.

Diversity produces more than monocultures. But monocultures are profitable to industry both for markets and political control. The shift from high-yielding diversity to low-yielding monocultures is possible because the resources destroyed are taken from the poor.

In contrast, the higher commodity production brings benefits to those with economic power. The polluter does not pay in industrial agriculture both of the chemical era or the biotechnological era. Ironically, while the poor go hungry, the hunger of the poor is used to justify the agricultural strategies that deepen their hunger.

Small Farms Are More Productive

The biotechnology industry often argues that only the industrial farms of the US can feed the world. It is falsely assumed that small farms and small farmers have low production. FAO's analysis has shown that small farms can be much more productive than large farms.

When one recognizes that the small farms of developing countries produce diverse outputs of nutritious crops, it becomes clear that industrial breeding has actually reduced food security by destroying small farms and small farmers. Protecting small farms is a food security imperative.

The displacement of small farms has been justified on grounds of alleged productivity of large farms. However, it is the small farms and small farmers who are being destroyed by globalization and trade-driven economic reforms.

The University of Essex, in the United Kingdom, completed an audit of progress towards agricultural sustainability in 208 projects in 52 developing countries (Pretty et al., 2002). These projects included both integrated, near-organic systems (179 cases), and certified and non-certified organic systems (29 cases). These organic cases comprised a mix of food, fiber, and beverage-based agriculture systems, with 154–742 households farming 106,197 hectares. The average area per household is small (0.7 ha), as many projects involve small-scale organic vegetable production. This audit indicated that promising improvements in food production are occurring through one or more mechanisms:

1. Intensification of a single component of the farm system, such as home-garden intensification with vegetables and trees
2. Addition of a new productive element to a farm system (such as fish in paddy rice) boosts the farm's total food production, income, or both, but does not necessarily affect cereal productivity
3. A better use of natural capital to increase total farm production, especially water

The diversity of crops on organic farms can have other economic benefits. Diversity provides some protection from adverse price changes in a single commodity. Diversified farming also provides a better seasonal distribution of inputs.

Organic farmers need to borrow less money than conventional farmers for two reasons. First, organic farmers buy less input such as fertilizer and pesticides. Second, costs and income are more evenly distributed throughout the year on diversified organic farms. Biodiverse organic farming based on agroecological principles addresses the above problems:

1. It leads to an increase in total farm productivity and increases net farm incomes by reducing costs of external inputs of seeds and chemicals to zero.

2. It creates a diversity of food distribution systems, from the household consumption of healthy food to the creation of local regional and national markets for real food from real farmers, trading only in high value, unique products that other regions cannot grow. Diversity of crops in our fields, and foods on our plates go hand in hand with global markets.
3. Pesticide and chemical-free production and processing brings safe and healthy food through fair trade to consumers at true costs.

Given the rapid changes in agriculture due to liberalization, there is an urgent need to monitor the ecological costs of globalization of agriculture using a biodiversity-based productivity framework to reflect the health of nature's economy and people's economy. We have developed the following framework over the past two decades:

A. Provides documentation of the biodiversity status of a farm
B. Indicates the contribution of biodiversity to provisioning of internal inputs and the building and maintenance of nature's economy through the conservation of soil, water, and biodiversity
C. Indicates the contribution of biodiversity to the self-provisioning of food needs by agricultural families and communities and the building and maintenance of people's economy
D. Reflects the market economy of the farm in terms of incomes from sale of agricultural produce, and of additional costs for the purchase of external inputs and food items, when benefits from biodiversity are foregone

Increasing Costs of Production

A primary reason for the agrarian crisis is the increasing cost of production and dependence on the purchases of external inputs, including fossil fuels, chemicals, and seeds. Seed monopoly has increased and with it the price of seed. Corporations promote a shift from open-pollinated seeds that farmers can save to hybrids, including GMO hybrids such as Bt cotton, so that farmers are forced to buy seed at high cost every year.

Chemical fertilizers and pesticides puts farmers on a chemical treadmill. The more farmers depend on chemicals, the more they must buy them because of declining response.

Ever since the advent of Green Revolution in 1960s, governments have adopted a policy to support chemical fertilizers through a subsidy system.

Chemical fertilizers are leading to a decline in productivity because they are destroying soil health.

Just as chemical fertilizer destroys soil ecology, pesticides, GMO plants destroy insect ecology and thus create more pests.

The primary justification for the genetic engineering of Bt into crops is that this will reduce the use of insecticides. The engineering of the genes for the Bt toxin implies that a high dose of this toxin is expressed in every cell of every plant all the time. Long-term exposure to Bt toxins promotes development of resistance in insect populations.

In the era of antibiotics, we similarly develop resistance to them. The problem with both the Bt and herbicide-tolerant (HT) crops is that they exert intense selection pressure on pest populations to evolve resistance, bringing on resistance much faster than would have otherwise occurred. Both Bt and HT crops are unsustainable technologies in agriculture.

For example, Bt cotton has contributed to the emergence of non-target pests such as ahids, jassids, army bug, mealy bug, and white fly. As a result, farmers are using more pesticides, bearing more expenses, and facing more frequent crop failure. The high costs and the risks are leading to increasing debt and suicides.

True Cost Misconceptions

Globalization of agriculture implies the corporate control of agriculture. The World Bank's structural adjustment policies forced the entry of global seed and grain corporations into India. The WTO Agreement on Agriculture forces countries to liberalize exports and imports, allowing global corporations to control domestic production, domestic markets, and global trade.

There are two major misconceptions about the myth of falling prices because of productivity and competition. The first is the incorrect definition of "productivity" in industrial agriculture, which is based on monocultures and the higher exploitation of nature and people.

The strategy is to treat our small peasants and small farms as an "obstruction" and imagine that corporatization of our agriculture and increases in size of farms will protect us. In fact, corporatization will destroy our farms and farmers by handing over our vital resources to TNCs, destroying our biodiversity which is the real capital in nature's economy, and destroying our small farmers which are the foundation of the food economy.

Imports threaten our survival externally, corporation is inviting the invader to take over our assets from within.

The second is the false assumption of competition based on promoting corporate control of agriculture, unnecessary exports, and imports under trade liberalization, hiding true costs through subsidies and externalities, putting corporate profits above farmer's lives and nature's sustainability.

Industrial agriculture is only sustained through subsidies to the agrichemical and agribusiness industry and the externalization of the cost on the environment and public health.

The patents on seed clauses in the Agreement on Trade Related Intellectual Property Rights (TRIPs) were drafted by Monsanto. Still, we were successful in excluding seeds, plants, and animals from WTO rules and from the Indian Patent Act through Art 3j. I was part of the expert group for drafting the Plant Variety Protection and Farmers Rights Act.

In addition, the Agreement of Sanitary and Phyto Sanitary Measures drafted by the junk food and industrial food processing industry—including Nestle, Pepsi, Coca-Cola—is forcing changes in Nation Food Safety laws, shutting down the small-scale artisanal sector and deregulating the industrial processing sector, based on chemicals and synthetic ingredients which require regulation to protect people's health.

WTO has encouraged seed monopolies through the TRIPS Agreement, as well as dumping subsidized food through the Agreement on Agriculture. These agreements have reduced food to "raw material" for industrial processing made possible by these pseudo-safety laws that prohibit local food while continuously deregulating industrially processed food.

Internal liberalization implies liberating agriculture to enhance self-regenerative ecological processes and enhance ecological and livelihood security. In particular this includes:

+ Freeing agriculture from high external inputs such as chemical fertilizers and pesticides and making a transition to sustainable agriculture based on internal inputs for ecological sustainability
+ Freeing farmers from capital intensive farming and debts
+ Freeing peasants from landlessness
+ Freeing farmers from fear of dispossession by monopolies of land, water, and biodiversity

+ Freeing people of unjust and coercive laws that criminalize local production, distribution, and processing go hand in hand with deregulation of corporations
+ Freeing the poor from starvation by ensuring food as a human right
+ Freeing people from unhealthy, toxic food which is spreading chronic diseases globally
+ Freeing rural people from water scarcity by ensuring inalienable and equitable water rights
+ Freeing knowledge and biodiversity from IPR monopolies
+ Rebuilding local food security while reinvigorating rural economies

The Economic Costs to Farmers, the Environment, and Public Health

Many people are bearing both high environmental and public health costs. Due to the chemicals in our food, industrial processing of our foods, and nutritionally empty commodities, many people are suffering from deficiencies of trace elements and micronutrients, which causes chronic diseases like diabetes and obesity.

The food we eat affects our health. Studies show that 51% of all food is contaminated with pesticides, including DDT. Pesticides mimic estrogen and oppose the actions of male hormones, which can lead to infertility.

The cancer epidemic is intensifying. Chemicals in agriculture and industry are causing this to happen. Navdanya, a nonprofit organization, ran a survey in Gangnauli village in the district of Baghpat which found that there are 100 patients suffering from cancer in that village alone.

Malnutrition in India is a crisis. One of the most common problems, food-related disease epidemics, affects nearly 39% of India's children. Lack of nutritious food is an issue for both the poor and the wealthy. Even amongst Indians who are better off, malnutrition is common.

The Green Revolution, forced on India by the US, removed all considerations for health and nutrition, and focused only on increasing the use of agrichemicals and the production of commodities. This resulted in increased production of nutritionally empty commodities, full of pesticides and toxins, and reduced the availability of nutritionally-rich foods.

Unhealthy food produced non-sustainably produces high financial costs to both nature's economy and the people's. Through this distortion of externalization of the cost of real food sold by farmers, the true costs is artificially made to look more expensive and less viable than real food actually is.

The real price of food leads to true cost accounting which internalizes the externalities and takes into account the heavy subsidies for industrial agroculture and the industrial processing industry.

The Navdanya study Wealth Per Acre (Table 9) has assessed the socio-ecological externalities of chemical farming in India to be $1.3 trillion annually (the fertilizer subsidies in India are $12.43 billion.)

Table 9: Total cost of all negative externalities

S. NO.	NEGATIVE EXTERNALITY	ANNUAL COST IN US$	% OF TOTAL OF ALL NEGATIVE EXTERNALITIES
1.	Farmers' Suicide	US$ 98.0 billion	7.8%
2.	Fatal poisoning	US$ 274.1 billion	21.7%
3.	Unspecific long-term diseases	US$ 32.9 billion	2.6%
4.	Specific long-term diseases	US$ 130.8 billion	10.4%
5.	Livestock animal poisoning	US$ 261.1 million	
6.	Companion animal poisoning	US$ 6.0 million	
7.	Wild animal poisoning	US$ 2.7 million	
8.	Bird poisoning	US$ 2.4 billion	3.3%
9.	Fish kills	US$ 36.4 million	
10.	Honeybee poisoning	US$ 349.7 million	
11.	Pesticide resistance and elimination of natural enemies	US$ 38.9 billion	
12.	Crop loss to industrial agriculture	US$ 69.7 billion	5.5%
13.	Surface and ground water contamination	US$ 45.0 billion	3.6%
14.	Air pollution	US$ 68.9 billion	5.5%
15.	Cost of loss of organic matter from soil	US$ 7.1 billion	0.6%
16.	Cost of loss of biodiversity in agroecosystems	US$ 467.5 billion	37.0%
17.	Cost of agricultural subsidies	US$ 27.4 billion	2.2%
	Total	**US$ 1.26 trillion/year**	**100%**

As we have shown in this book, industrial farming has high ecological costs which are destroying the very conditions of life. The high costs have driven farmers to suicide. The high costs of chemical and junk food are adding up to an unattainable price for industrial food.

Food transported long distances requires processing, lots of chemical treatment, refrigeration, and packaging that directly contributes to pollution, diseases, and climate change. All of this packaging ends up as mountains of garbage near or in our cities. Greenhouse gases, such as carbon dioxide from "food miles" and methane from garbage dumps, are contributing to climate change and destabilizing the planet.

Eating local and creating a sustainable and healthy foodshed for your city means reducing food miles and toxins in the food chain. Eating local means we are connecting directly with our farmers and helping them shift to agriculture that allows them to grow biodiverse, safe, healthy food that we can have access to.

Rebuilding the broken food system, its ecological cycles and the broken links between the city and the countryside, means creating food-smart citizens who know what they are eating and where their food comes from.

Farmers have been committing suicide because they are spending too much on chemicals and seeds and do not receive a fair price for what they produce through their hard work.

By ensuring that a fair share of what consumers spend reaches our food providers we can help end farmers' suicides. We can rejuvenate our health while rejuvenating the agricultural economy and the Earth.

The industrial, corporate, globalized model of agriculture is based on the production of a few globally traded commodities with high external inputs. Thus farmers growing Bt cotton spend more than they earn and are pushed into debt, and in extreme cases to suicide.

On the other hand, biodiversity intensification with native seeds increases both output and incomes.

Monocultures do not benefit the farmer. They may produce more for corporations, but they create the dependency of farmers on their monopoly markets and high cost inputs. Monocultures also contribute to increasing profits of corporations, which sell costly inputs.

Organic Agriculture Creates Higher Net Incomes Worldwide

Organic product sales, the number of hectares under organic production, as well as the number of organic producers continues to increase despite the current difficult global economic conditions. Information from most countries around the world is showing a consistent trend of a dynamic and growing industry. As with all trends, there are examples of fluctuations in the rate of growth and even limited periods in some countries where there can be small declines; however, the meta data shows a strong active positive trend. Despite the mixtures of economic downturns, sluggish growth, and other market uncertainties in countries around the world, the organic sector continues to grow and outperform most other agri-food sectors.

The drivers of organic growth globally can be divided into two main categories: the pull from consumers and markets and the push from resilient production systems. Initially the growth in the organic movement was driven by farmers from the 1920s to the 1980s who were concerned about the loss of crop quality and economic viability due to the use of synthetic chemical fertilizers and pesticides. The resilience and appropriateness of organic systems, especially in terms of adaptation to climate extremes and providing food security, continues to be a critical trend in the sustainable growth in the organic sector.

The consumer-based market pull is a considerable driver of growth in the organic sector, especially with the emergence of the third-party certified sector in the 1980s. Organic guarantee systems, such as the USDA National Organic Program certification system, are designed to ensure that consumers can trust the integrity of organic products and consequently have an important role driving demand.

Two other global trends are having an important influence on the growth of the sector. Despite the constant worldwide decline in the number of farmers, there are steady increases in the numbers of organic producers due to their economic viability. Another important trend is the beginning of a science and research-based approach after almost a century of being largely ignored.

The publication of *Silent Spring* in 1962 raised the issue of toxic chemicals in food and in the environment. This was the beginning of the organic consumer movement due to their concerns over toxic chemicals in food. The concern over pesticides is still the main consumer driver. Surveys by

Newspoll (OFA 2008), Nielsen, Organic Trade Association, and other credible organizations show that over 60% of main grocery buyers in Australia and Canada, 78% in the USA, and substantial percentages in EU countries make some purchases of organic products, showing that there is a high level of recognition and acceptance of organic products globally.

This research shows that the increase in global demand for organic food is being driven by consumers who are concerned about health, the environment, and food quality. The primary reason for choosing organic food is due to health concerns associated with toxic chemicals, followed by the belief that organic foods are more nutritious.

This is part of the growing trend worldwide for healthier and safer foods. This rapidly growing market segment is called LOHAS (Lifestyles of Health and Sustainability). The organic sector has the largest dedicated and best-known market share.

A viable income is an essential part of farm sustainability. The most recent study by Noémi Nemes from the United Nations Food and Agriculture Organization (FAO) analyzed of over 50 economic studies. She stated that the data:

> "[D]emonstrates that, in the majority of cases, organic systems are more profitable than non-organic systems. Higher market prices and premiums, or lower production costs, or a combination of the two generally result in higher relative profits from organic agriculture in developed countries. The same conclusion can be drawn from studies in developing countries, but there, higher yields combined with high premiums are the underlying causes of their relatively greater profitability."

A report by United Nations Environmental Program and United Nations Conference on Trade and Development found that not only did organic production increase the amount of food production it also gave farmers access to premium value markets.

A study by Iowa State University in the US found that cost-wise, on average, the organic crops' revenue was twice that of conventional crops due to the savings from non-utilization of chemical fertilizers and pesticides.

A study in the US by Dr. Rick Welsh of the Wallace Institute has shown that organic farms can be more profitable. The premium paid for organic produce is not always a factor in this extra profitability. Dr. Welsh analyzed a diverse set of academic studies comparing organic and conventional cropping systems. Among the data reviewed were six university studies that compared organic and conventional systems (Welsh 1999).

Chemical farming is trapping farmers in debt and giving corporations additional profits by extracting nutrition from food through industrial processing. Both producers and eaters lose, while corporations make profits from both.

However, the economic paradigm based on a linear one-way extraction of resources and wealth from nature and society has promoted systems of production and consumption that have ruptured and torn apart ecological cycles, threatening the stability of the natural and social world. The extractive model in agriculture is the root cause of the ecological crisis, the agrarian crisis, and the disease epidemic.

The data shows that organic agriculture is the future of farming due to fast growing high value, premium markets, strong consumer demand, and high yielding, resilient, productions systems. The organic sector covers the whole value chain from the seed to the table, and from field to the plate; consequently, it presents numerous opportunities for all types of investors, from the smallest family farmers and retailers through to the largest corporations. As long as there are good management practices, all will prosper.

Industrial agriculture is based on ecological and social unsustainability. A transition to ecological agriculture and just food systems has become an imperative to regenerate nature's economy and people's economy.

Biodiversity-based organic farming and biodiversity of markets and economies is Navdanya's approach to rejuvenate soil, water and biodiversity, rural economies, and the health of all through cooperation, circular economies, and local food systems.

Introduction

A Viable Food Future, Part I, (2010), The Development Fund/Utviklingsfondet, Norway.

Aldridge, J., Seidler, F., Meyer, A., Thillai, I., and Slotkin, T. (2003), Serotonergic Systems Targeted by Developmental Exposure to Chlorpyrifos: Effects during Different Critical Periods, Environmental Health Perspectives Vol. 111, Number 14, November 2003.

Agriculture and Food Policy Reference Group (2006), Creating Our Future: Agriculture and Food Policy for the Next Generation, Report to the Minister for Agriculture, Fisheries and Forestry, Canberra, February, ISBN 1 920925 49 X Published by ABARE, GPO Box 1563 Canberra ACT 2601.

Anderson, I., (2010), Agricultural Development, Food Security and Climate Change: Intersecting at a global Crossroads,http://web.worldbank.org/Wbsite/External/Topics/Extsdnet/0,,content MDK:22782615~pagePK:64885161~piPK:64884432~theSitePK:5929282,00.html

Avery, D. (2000), Saving the Planet with Pesticides and Plastic: The Environmental Triumph of High-Yield Farming, Hudson Institute, USA.

Azeez, G. (2009), Soil Carbon and Organic Farming, Soil Association, Bristol, UK, November (2009), http://www.soilassociation.org/Whyorganic/Climatefriendly-foodandfarming/Soilcarbon/tabid/574/Default.aspx 2010, ISBN 978-82-91923-19-2 (Printed edition) ISBN 978-82-91923-20-8 (Digital edition).

Badgley et al., (2007), Organic agriculture and the global food supply, Renewable Agriculture and Food Systems (2007), 22: 86-108 Cambridge University Press doi:10.1017/S1742170507001640.

Bardgett, R. D. (2005), The biology of soil: a community and ecosystem approach, Oxford University Press Inc, New York.

Barrett, C. et al., (2004), Better technology, better plots, or better farmers? Identifying changes in productivity and risk among Malagasy rice farmers Am.J.Agric.Econ., 2004, 86, 4, 869-888.

Bell, P. (1992), Eutrophication and coral reefs: some examples in the Great Barrier Reef lagoon, Water Research, 26: 553-568.

Bell, P., Elmetri, I., (1995), Ecological Indicators of Large-scal Eutrophication in the Great Barrier Reef Lagoon, Ambio (1995) Vol. 24, Issue: 4, pp. 208-215.

Bemani, J. M. et al., (2005), Agricultural run-off fuels large phytoplankton blooms in vulnerable areas of the ocean, Nature 434, 211-214 (10 March 2005); doi:10.1038/nature03370.

Benbrook, C. M. (2005), Second State of Science Review (SSR), Organic Centre, USA www.organic-center.org/stateofscience.htm

Buznikov, G. A., et al., (2001), An Invertebrate Model of the Developmental Neurotoxicity of Insecticides: Effects of Chlorpyrifos and Dieldrin in Sea Urchin Embryos
and Larvae, Environmental Health Perspectives Vol. 109, Number 7, July 2001.

Cabello, G., et al., (2001), A Rat Mammary Tumor Model Induced by the Organo-
phosphorous Pesticides Parathion and Malathion, Possibly through Acetylcho-
linesterase Inhibition, Environmental Health Perspectives Vol. 109, Number 5,
May 2001.

Cacek, T. and Langner L. L. (1986), The economic implications of organic farming,
1986, American Journal of Alternative Agriculture, Vol. 1, No. 1, pp. 25-29.

Cadbury, D. (1997), The Feminization of Nature, Penguin Books, Middlesex England, 1998.

Carson, R. (1962), Silent Spring, Penguin Books, New York, USA 1962.

Cavieres, M., Jaeger, J. and Porter, W. (2002), Developmental Toxicity of a Commercial Herbicide Mixture in Mice: I. Effects on Embryo Implantation and Litter
Size, Environmental Health Perspectives Vol. 110, Number 11, November 2002.

Charizopoulos, E. and Papadopoulou-Mourkidou, E. (1999), "Occurrence of Pesticides in Rain of the Axios River Basin, Greece," Environmental Science & Technology [ES&T] Vol. 33, No. 14 (July 15, 1999), pp. 2363-2368.

Clements, C., Ralph S. and Petras M. (1997).Genotoxicity of select herbicides in
Ranacatesbeiana tadpoles using the alkaline single cell gel DNA electrophoresis
(comet) assay. Environ. Mole. Mutagen. 29:277- 288.45.

Colborn, T., Dumanoski, D. and Myers J. P., (1996) Our Stolen Future, March 1996.

Connor, S. (2011), The Independent (Major UK Newspaper) 22 October 2011.
Reporting on the scientific study that is in the process of being published in a
peer reviewed journal, *Geophysical Research Letters.*

Cox, C. (2004), Glyphosate (Roundup) Journal Of Pesticide Reform, Winter, Vol.
24, No. 4, Northwest Coalition Against Pesticides, Eugene, Oregon.

Cox, C. (2001), Atrazine: Environmental Contamination and Ecological Effects,
Journal Of Pesticide Reform, Fall 2001, Vol. 21, No. 3, p12, Northwest Coalition
Against Pesticides, Eugene, Oregon.

Drinkwater, L. E., Wagoner, P. and Sarrantonio, M. (1998), Legume-based cropping
systems have reduced carbon and nitrogen losses. Nature 396, 262-265 (1998).

ETC Group 2009, Who Will Feed Us? Questions for the Food and Climate Crises,
ETC Group, November 2009, www.etcgroup.org

FAO (2000), Twenty Second FAO Regional Conference for Europe, Porto, Portu-
gal, 24-28 July 2000 Agenda Item 10.1, Food Safety and Quality as Affected by
Organic Farming.

FAO (2003), Organic Agriculture: the Challenge of Sustaining Food Production
while Enhancing Biodiversity, Food and Agriculture Organization of the United
Nations, Rome, Italy.

FAO (2007), FAO international conference on Organic Agriculture and Food Security, 2007: Organic agriculture can contribute to food security, Food and Agriculture Organization of the United Nations, Rome, Italy

FAO (2010), The State of Food Insecurity in the World, Food and Agriculture Organization of the United Nations, Rome, Italy.

Garry, V. F., et al., (2001), Biomarker Correlations of Urinary 2,4-D Levels in Foresters: Genomic Instability and Endocrine Disruption, Environmental Health Perspectives Vol. 109, Number 5, May 2001.

Handrek, K. (1990), Organic Matter and Soils, CSIRO, Australia, 1979, reprinted 1990.

Handrek, K. and Black, N. (2002) Growing Media for Ornamental Plants and Turf, UNSW Press, Sydney 2002.

Hardell, L. and Eriksson M. (1999), "A Case-Control Study of Non-Hodgkin Lymphoma and exposure to Pesticides," Cancer Vol. 85, No. 6 (March 15, 1999), pp. 1353-1360.

Hayes, T. B., et al., (2002). "Hermaphroditic, demasculinized frogs after exposure to the herbicide atrazine at low ecologically relevant doses." Proceedings of the National Academy of Sciences, Vol. 99:5476-5480, April 16, 2002.

Hayes, T. B., et al., (2003), Atrazine-Induced Hermaphroditism at 0.1 ppb in American Leopard Frogs (Rana pipiens): Laboratory and Field Evidence Environmental Health Perspectives Vol. 111, Number 4, April 2003.

Hoegh-Guldberg, O., Mumby, P.J., Hooten, A. J., Steneck, R. S., Greenfield, P., Gomez, E., Harvell, E.D., Sale, P.F., Edwards, A. J., Caldeira, K., Knowlton, N., Eakin, C. M., Iglesias-Prieto, R., Muthiga, N., Bradbury, R. H., Dubi, A. and Hatziolos, M. E., (2007), Coral Reefs Under Rapid Climate Change and Ocean Acidification, Science 318: 1737-1742.

Hole, D., Perkins A, Wilson J, Alexander I, Grice P. and Evans A., (2004) Does organic farming benefit biodiversity?, Biological Conservation Vol. 122, Issue I pp. 113–130.

Howe, C.M. et al., (2004), Toxicity of glyphosate based pesticides to four North American frog species. Environ. Toxicol. Chem. 23:1928-1938.

Huber, D. M. (2010), Ag Chemical And Crop Nutrient Interactions, Fluid Journal, Spring 2010, Vol. 18 No. 3, Issue #69.

IAASTD (2008), International Assessment of Agricultural Knowledge, Science and Technology for Development (IAASTD), Island Press, 1718 Connecticut Avenue NW, Suite 300 Washington DC, 20009-1148 info@islandpress.org

IEA (2011), Prospect of limiting the global increase in temperature to 2°C getting bleaker, International Energy Agency Media Release 30 May 2011, http://www.iea.org/index_info.asp?id=1959.

IFOAM (2011), IFOAM Smallholder Position Paper in publication www.ifoam.org

Immig J, (2010), A list of Australia's most Dangerous Pesticides, a report by the National Toxics Network and the World Wildlife Fund. http://ntn.org.au/2011/01/10/toxic-hit-list-shows-australians-exposed-to-dangerous-pesticides/

Infopest (2004), Queensland Department of Primary Industries and Fisheries. Primary Industries Building, 80 Ann St, Brisbane, Queensland, Australia.

International Agency for Research on Cancer (1999), "Overall Evaluations of Carcinogenicity to Humans 6-Chloro-N-ethyl-N¢-(1-methylethyl)-1,3,5-triazine-2,4-diamine" Vol. 73 (1999) (p. 59).

Jones, C. E. (2006), Balancing the Greenhouse Equation–Part IV, Potential for high returns from more soil carbon, Australian Farm Journal, February 2006, pp. 55-58.

Jones, C. E. (2011), Carbon that Counts, in publication. A preview text is available at www.ofa.org

Khan, S. A., Mulvaney, R. L., Ellsworth, T. R., and Boast (2007), C. W. The Myth of Nitrogen Fertilization for Soil Carbon Sequestration. Journal of Environmental Quality. 2007 Oct 24; 36(6):1821-1832.

Lajmanovich, R. C., Sandoval, M. T. and Peltzer P. M. (2003). Induction of mortality and malformation in Scinax nasicus tadpoles exposed to glyphosate formulations. Bull. Environ. Contam.Toxicol. 70:612-618.46

Lal, R. (2008), Sequestration of atmospheric CO_2 in global carbon pools, Energy and Environmental Science 1: 86–100. doi:10.1039/b809492f.

Lal, R. (2007), Carbon sequestration Phil. Trans. R. Soc. B 27 February 2008 Vol. 363 No. 1492 815-830 doi: 10.1098/rstb.2007.2185.

LaSalle, T. and Hepperly, P. (2008), Regenerative Organic Farming: A Solution to Global Warming, The Rodale Institute 611 Siegfriedale Road Kutztown, PA 19530-9320 USA.

Leu, A. F. (2004), Organic Agriculture Can Feed the World, Acres USA, Vol. 34, No. 1

Lotter, D. W., Seidel, R. and Liebhart, W. (2003), The performance of organic and conventional cropping systems in an extreme climate year. American Journal of Alternative Agriculture 18(3):146–154.

Mader, P., Fliessbach, A., Dubois, D., Gunst, L., Fried, P. and Niggli, U. (2002), Soil fertility and biodiversity in Organic Farming. Science 296, 1694-1697.

MA Report (2005), Millennium Ecosystem Assessment Synthesis Report, The United Nations Environment Programme March 2005.

MBDA (2011) The official website of the Murray Darling Basin Authority, http://www.mdba.gov.au/water/blue-green-algae.

Mills, P. et al. (2002), Cancer Incidence in the United Farmworkers of America (UFW) 1987-1997, 2001, American Journal of Industrial Medicine, 40: pp. 596–603, 2002.

Monbiot G (2000), Organic Farming Will Feed the World, Guardian, 24th August 2000.

Mulvaney R. L., Khan, S. A. and Ellsworth, T. R., (2009), Synthetic Nitrogen Fertilizers Deplete Soil Nitrogen: A Global Dilemma for Sustainable Cereal Production, Journal of Environmental Quality 38:2295-2314 (2009) doi: 10.2134/jeq2008.0527, American Society of Agronomy, Crop Science Society of America, and Soil Science Society of America 677 S. Segoe Rd., Madison, WI 53711 USA.

National Standard (2005), National Standard for Organic and Bio-Dynamic Produce, *Edition 3.1, As Amended January 2005*, Organic Industry Export Consultative

Committee, c/o Australian Quarantine and Inspection Service, GPO Box 858, Canberra, ACT, 2601.

New Scientist (2001), Editorial, February 3, 2001.

Nordstrom, M. *et al.*, (1998), "Occupational exposures, animal exposure, and smoking as risk factors for hairy cell leukaemia evaluated in a case-control study," British Journal Of Cancer Vol. 77 (1998), pp. 2048-2052.

OFA (2008), The Newspoll survey commissioned by the Organic Federation of Australia can be downloaded at: http://www.ofa.org.au/papers/OFA_Newspoll_Report_2008.pdf

Organic Production Survey (2008), Vol. 3, Special Studies, Part 2, AC-07-SS-2, Issued February 2010, Updated July 2010, United States Department of Agriculture, http://www.agcensus.usda.gov/Publications/2007/Online_Highlights/Organics/

Parrott, N., (2002), "The Real Green Revolution," Greenpeace Environmental Trust, Canonbury Villas, London ISBN 1 903907 02 0.

Pearce, F. and Mackenzie, D., (1999), "It's raining pesticides; The water falling from our skies is unfit to drink," New Scientist April 3, 1999, p. 23.

Pimentel, D. et al., (2005), Environmental, Energetic and Economic Comparisons of Organic and Conventional Farming Systems, Bioscience (Vol. 55:7), July 2005.

Porter, W. et al., (1999), "Endocrine, immune and behavioral effects of aldicarb (carbamate), atrazine (triazine) and nitrate (fertilizer) mixtures at groundwater concentrations," Toxicology and Industrial Health (1999) 15, 133-150.

Posner et al., (2008), Organic and Conventional Production Systems in the Wisconsin Integrated Cropping.... Agron J.2008; 100: 253-260.

Prange, J. et al., (2007), Great Barrier Reef Water Quality Protection PlanAnnual Marine Monitoring Report, Reporting on data available from December 2004 to April 2006, Great Barrier Reef Marine Park Authority 2007 ISSN 1832-9225.

Pretty, J. (1998a), The Living Land-Agriculture, Food and Community Regeneration in Rural Europe, Earthscan, London. August 1999.

Pretty, J. (1998b), SPLICE magazine, August/September 1998 Vol. 4 Issue 6.

Pretty, J. (1995), Regenerating Agriculture: Policies and Practice for Sustainability and Self-Reliance, Earthscan, London.

Qiao, D. et al., (2001), Developmental Neurotoxicity of Chlorpyrifos Modeled in Vitro: Comparative Effects of Metabolites and Other Cholinesterase Inhibitors on DNA Synthesis in PC12 and C6 Cells, Environmental Health Perspectives Vol. 109, Number 9, Sept. 2001.

Ravishankara, A. R., Daniel, J. S. and Portmann, R.W. (2009), "Nitrous Oxide (N_2O): The Dominant Ozone-Depleting Substance Emitted in the 21st Century". Science 326 (5949).

Reganold, J., Elliott, L. and Unger, Y., (1987), Long-term effects of organic and conventional farming on soil erosion, Nature 330, 370-372 (26 November 1987); doi:10.1038/330370a0.

Reganold, J., Glover, J., Andrews, P. and Hinman, H. (2001), Sustainability of three apple production systems. Nature 410, 926-930 (19 April 2001) | doi: 10.1038 /35073574123–5. doi:10.1126/science.1176985.

Reuben, S. H., for The President's Cancer Panel, (2010), Reducing Environmental Cancer Risk What We Can Do Now, 2008–2009 Annual Report i President's Cancer Panel, April 2010, U.S. Departmenof Health and Human Services, National Institutes of Health, National Cancer Institute.

Rodale (2011), http://www.rodaleinstitute.org/about_us

Rodale (2003), Farm Systems Trial, The Rodale Institute 611 Siegfriedale Road Kutztown, PA 19530-9320 USA

Rodale (2006), No-Till Revolution, The Rodale Institute 611 Siegfriedale Road Kutztown, PA 19530-9320 USA http://www.rodaleinstitute.org/no-till_revolution.

Sala, O. E., Meyerson, L. A., Parmesan, C. (26 January 2009). Biodiversity change and human health: from ecosystem services to spread of disease. Island Press. pp. 3–5. ISBN 9781597264976.

Sanderman, J., Farquharson, R. and Baldock, J., Soil Carbon Sequestration Potential: A review for Australian agriculture, CSIRO Land and Water, 2010.

Steer, A. 2011, Agriculture, Food Security and Climate Change-A Triple Win? http://site resources.worldbank.org/INTSDNET/64884474-1244582297847/ 22752519/Hague_Opening_F.pdf

Stevenson, J. (1998), Humus Chemistry in Soil Chemistry p. 148 Wiley Pub. NY 1998.

Short, K. (1994), Quick Poison, Slow Poison, 1994, ISBN 0 858811278.

Steingraber, S. (1997), Living Downstream; An Ecologist Looks At Cancer And The Environment, New York: Addison-Wesley, 1997.

Storrs, S. et al., (2004), Survivorship Patterns of Larval Amphibians Exposed to Low Concentrations of Atrazine, Environmental Health Perspectives 112: No. 10.1054-1057 (2004).

Teasdale, J. R., Coffman, C. B. and Mangum, R. W. (2007), Potential long-term benefits of no-tillage and organic cropping systems for grain production and soil improvement. Agron. J. 99:1297-1305.

Thieu, V., Billen, G., Garnier, J., and Benoý^t Marc, (2010), Nitrogen cycling in a hypothetical scenario of generalised organic agriculture in the Seine, Somme and Scheldt watersheds. Reg Environ Change (2011), 11:359–370 doi: 10.1007/ s10113-010-0142-4, Published online: August 2010. This article is published with open access at Springerlink.com.

Thieu, V., Billen, G., and Garnier, J., (2009), Assessing the effect of nutrient mitigation measures in the watersheds of the Southern Bight of the North Sea, Science of the Total Environment, Published online: Jan 2010, This article is published with open access at Springerlink.com

Tilman, D., Fargione, F., Wolff, B., D'Antonio, C., Dobson, A., Howarth, R., Schindler, D., Schlesinger, W., Simberloff, D., and Swackhamer, D. (2001), Forecasting Agriculturally Driven Global Environmental Change, Science 13 April 2001 292: 281-284 [doi: 10.1126/science.1057544].

Trewavas, A. J. (2001), Urban myths of organic farming. Nature 410, pp. 409-410.

Unep-Unctad (2008), Organic Agriculture and Food Security in Africa, Sept 2008. http://www.unep-unctad.org/cbtf/index.htm

UNFCCC (2011), The United Nations Framework Convention on Climate Change, http://cancun.unfccc.int/

USEPA (2002), United States Environmental Protection Agency Revised Human Health Risk Assessment Atrazine April 16, 2002 Reregistration Branch 3 Health Effects Division Office of Pesticide Programs

Welsh, R. (1999), Henry A. Wallace Institute, The Economics of Organic Grain and Soybean Production in the Midwestern United States, Policy Studies Report No. 13, May 1999.

Willer, H. and Kilcher, L. (Eds.) (2011), The World of Organic Agriculture-Statistics and Emerging Trends 2011. IFOAM, Bonn, and FiBL, Frick.

Wynen, E. (2006), 'Economic management in organic agriculture'. In: Kristiansen, P., Taji, A. and Reganold, J. P. (Eds): Organic Agriculture-a Global Perspective, Chapter 8, CSIRO Publishing, Melbourne (see chapter)

Sections 2-7

Aggarwal, P. K. (2003), "Impact of Climate Change on Indian Agriculture" J Plant Biol., Vol. 30, 2. Conway, Gordon 1997. The Doublyst Green Revolution: First of All in the 21st Century. London. Penguin.

Altieri, M. A., (1993), Biodiversity and Pest Management in Agroecosystems. New York: Food Products Press.

Altieri, M. A. (1987), Agroecology: The Scientific Basis of Alternative Agriculture. Boulder: Westview Press.

Altieri, M. A. (2000), Agro-ecology: Principles and strategies for designing sustainable farming systems. Agroecology in Action.

Bhatt, V. K., Singh, H. R. and Semwal, M. M. (2007), Biodiversity Based Organic Agriculture: A sustainable Livelihood option for Uttarakhand, Samaj Vigyan Shodh Patrika, Special Issue (Uttarakhand -1), pp. 65-79.

Deb, D. (2004), Industrial vs Ecological Agricultural, Navdanya/Research Foundation for Science technology and Ecology, New Delhi, p. 80.

Deere, C. D., (1997), Reforming Cuban Agriculture. Development and Change 28, 649-669.

DEFRA (2005), "Climate Change Impacts on Agriculture in India" A Joint Program Ministry of Environment and Forest, Govt. of India and Department of Environment, Food and rural Affairs (DEFRA), UK, Sept. 2005.

Dobhal, D. P., Gergan, J.T., Thayyen, R.J. (2004), Recession and Morphogeometrical changes of Dokriani glacier (1962-1995), Garhwal Himalaya, India. Current Science, 86 (5), pp. 101-107.

Dobhal, D. P., Gergan, J. T. and Thayyen, R. J. (2007), Recession and Mass balance fluctuations of Dokriani glacier from 1991 to 2000, Garhwal Himalaya, India, In:

International seminar "Climatic and Anthropogenic impacts on water resources variability"

Dobhal, D. P., Gergan, J. T. and Thayyen, R. J. (2008), Mass balance studies of the Dokriani Glacier from to, Garhwal Himalaya, India, Bulletin of Glaciological Research (25) 9-17.

Dulal, Goswami, in theshillongtimes.com, 27-Aug, 2007. Indian Institute of Tropical Meteorology (IITM), Pune. Funes-Monzote, R., 2004. De bosque a sabana. Azúcar, deforestación y medioambiente en Cuba, 1492-1926, Siglo XXI Editores, México D.F.

Funes, F., García, L., Bourque, M., Pérez, N., Rosset, P., (2002), Sustainable agriculture and resistance. Transforming Food Production in Cuba. Food First books, Oakland.

FAO (Food and Agriculture Organization of the United Nations). (2000), State of the World's forests 1997. FAO, Rome, Italy.

Fujisaka, S. 1999. *Side-stepped by the Green Revolution: Farmers' traditional rice cultivars in the uplands and rainfed lowlands.* pp. 50-63. In: Prain, G., Fujisaka, S. and Warren, M. D. (eds.), Biological and Cultural Diversity: The role of indigenous agriculture experimentation in development. London. Intermediate Technology Publications.

Grabherr, G., Gottfriend, M., Pauli, H. (1994), *'Climate Effects on Mountain Plants'.* In Nature, 369: 448. *Hasnain, S. I., 2002: Himalayan glaciers meltdown: impact on South Asian Rivers. IAHS Pub No. 274, pp. 1-7.* Howard, Louise E, 1953. "Sir Albert Howard, Soil and Health", Faber and Faber, London, p. 15.

Howard, A., (1940), An Agriculture Testament, London http://nature.berkeley.edu/~miguel-alt/principles_and_strategies.html (retrieved on December 2, 2013).

ICIMOD/UNEP (2001), *Inventory of glaciers, glacier lakes and glicial lake outburst floods, monitoring and early warning system in the Hindu Kush-Himalayan Region. Nepal,* p. 247. ICIMOD/UNEP, Kathmandu.

ICIMOD, 5 June 2007, Ms Bidya Banmali Pradhan, http://www.roap.unep.org/press/NR07-10.html.

Imperils the sustainability of the agricultural systems in the Central Himalaya. Curr. Sci. 9: 771-782 Marrero, L., 1974-1984. Cuba, Economía y Sociedad, Play- or, Madrid. (15 volumes) McCully, Patrick, International Rivers Network, Press Release, May 9, 2007.

India's dams largest methane emmiters among the world's dams, May 18, 2007, http://www.icrindia.org/?p=146.

International Assessment of Agricultural Knowledge, Science and Technology for Development (IAASTD, 2009), Executive Summary of the Synthesis Report, Island Press, Washington, DC, USA

International *Hydrological Programme (IHP)-VI, UNE- SCO, Tech. Document,* 80, pp. 53-63.

IPCC, (2001), http://www1.ipcc.ch/pdf/climate-changes-2001/synthesis-spm/synthesis-spm-en.pdf.

IPCC (2007), IPCC Summary for Policymakers: Climate Change 2007: Climate Change Impacts, Adaptation and Vulnerability.

IPCC WGII Fourth Assessment Report. Jackson, Michael (1995), *Protecting the Heritage of rice biodiversity.* Geojournal 35: 267-274. Jangpangi, B. S., *Study of some Central Himalayan glaciers. J. Sci. Ind. Res.*, 1958, 17, 91–93.

Kadota, J. I., Sakito, O., Kohno, S., Sawa, H., Mukae, H., Oda, H., Kawakami, K., Fukushima, K., Hiratani, K. and Hara, K. (1993) *A mechanism of erythromycin treatment in patients with diffuse panbronchiolitis. Am. Rev. Respir. Dis.*, 147, pp. 153-159.

Kaul, M. K. (ed.), *Inventory of the Himalayan Glaciers*, GSI Special, Publication No. 34, 1999, p. 10.

Kaul, M. K. (1999), *Inventory of the Himalayan Glaciers: A Contribution to the International Hydrological Programme.* edited by Kaul, M. K. p. 165.

Kulkarni, A. V., Bahuguna, I. M., Rathore, B. P., Singh, S. K., Randhawa, S. S., Sood, R. K. and Dhar S., (2007), *Glacial retreat in Himalaya using Indian Remote Sensing Satellite data*, Curr. Sci., Vol. 92, No. 1, 10 January, 2007.

Le Riverend, J., (1970), Historia económica de Cuba, Instituto Cubano del Libro, Havana.

Lima, I. B. T., Ramos, F. M., Bambace, L. A. W. and Rosa, R. (2008) *Methane Emissions from Large Dams as Renewable Energy Resources: A Developing Nation Perspective*, Mitig Adapt Strat Glob Change (2008) 13:193–206 (published online 2007, http://tinyurl.com/2bzawj).

Longstaff, T. G. (1910), *Glacier Exploration in the Eastern Karakoram*, Geographical Jour. V. 35, pp. 622-658. Maikhuri, R. K, Semwal, R. L, Rao, K. S., Nautiyal, S. and Saxena, K. G. 1997. Eroding traditional crop diversity

MEA. (2005), Millennium Ecosystem Assessment: Ecosystems and Human Well-Being: Synthesis. Washington: Island Press. http://www.millenniumassessment. org/ documents/document.429.aspx.pdf.

Ministry of Environment and Forests (2004), *"India's National Communication to the United Nations Framework Convention on Climate Change"* M O E. & F, Govt. of India 2004, New Delhi.

Moreno, F. M., (1978), El Ingenio. Complejo económico social cubano del azúcar. Ciencias Sociales, Havana.

Naithani, A. K., Nainwal, H. C. and Prasad, C. P., Geomorphological evidences of retreat of Gangotri glacier and its characteristics. Curr. Sci., 2001, 80, 87–94.

National Water Policy 2002. http://wrmin.nic.in/writere-addata/linkimages/nwp 20025617515534.pdf

Oberoi, L. K., Siddiqui, M. A. and Srivastava, D. (2001). *Recession of Chipa, Meola and Jhulang (Kharsa) Glaciers in Dhauliganga Valley between 1912 and 2000.* GSI Special Publication, 11(65): 57–60.

Olivier De S. and Gaëtan, V. (2011), The New Green Revolution: How Twenty-First-Century Science Can Feed the World http://www.thesolutions- journal.com/ node/971), Vol. 2: Issue 4: Aug 18.

Parmesan, C., Yohe, G. (2003), 'A Globally Coherent Fingerprint of Climate Change Impacts Across Natural Systems'. In Nature, 421: 3742.

Pebley, A. R., (1998), *Demography and the environment. Demography*, 35 pp. 377-389.

Pérez, R. N., Echeverría, D., González, E., García, M., (1999), Cambios tecnológicos, sustentabilidad y participación. Universidad de La Habana, Havana, p. 273.

Pounds, J. A., Fogden, M. P. L. and Campbell, J. H. (1999), Ecology: Clouded futures. Nature, 398: 611-615. Qin Dahe, P. A. Mayewski, C. P. Wake, Kang Shichang, Ren Jiawen, Hou Shugui, Yao Tandong, Yang Qinzhao, Jin Zhefan, Mi Desheng, 2000. *Evidence for recent climate change from ice cores in the central Himalaya. Annals of Glaciology*, 31: 153-158.

Raina, V. K., (2004), Is the *Gangotri Glacier receding at an alarming rate? J. Geol. Soc. India*, 64, 819–821. Reijntjes, C., Haverkort, B. and Water-Bayer, A. 1992. *Farming for the Future: An Introduction to Low-External-Input and Sustainable Agriculture.* London: MacMillan Press. p. 250.

Richharia, R. H. and Goapalswami, S., (1990), *Rices of India*, Academy of Development Science, Kashele, Maha-rashtra, p. 350.

Rosset, P., Benjamin, M., (1994), The greening of the revolution: Cuba's experiment with organic agriculture. Ocean Press, Melbourne.

Rupa Kumar, K., Krishna Kumar, K., Prasanna, V., Kamala, K., Deshpande, N. R., Patwardhan, S. K. and Pant, G. B. (2003), Future Climatic Scenario. In: Shukla, P. R., Sharma, Subodh, K., Ravindranath, N. H., Garg, A. and Bhattacharya, S., (Eds.) *Climatic Change and India: Vulnerability Assessment and Adaptation.* Universities Press (India) Pvt. Ltd., Hydrabad.

Shah, T., (1993), Agriculture and rural development in 1990s and beyond. Economic and Political Weekly (September 25); A 74-A76.

Shiva, V., (2001), *The Violence of the Green Revolution*. Penang. Third World Network.

Shiva, V., (2008), Soil not Oil: Environmental Justice in an age of Climate Crisis, South End Press, Cambridge, Massachusetts, p. 147.

Shiva, V., (1991), *The Violence of the Green Revolution: Third World Agriculture, Ecology and Politics*. Penang. Third World Network.

Shiva, V., (2008), Soil Not Oil-Climate change Peak Oil and Food Insecurity, Women unlimited, New Delhi, p. 156.

Shiva, V., and Poonam, P., (2006), Biodiversity Based Organic Farming: A new Paradigm for Food Security and Food Safety, Navdanya, New Delhi.

Shiva, V., and Singh, V. (2011), Health Per Acre: Organic Solution to Hunger and Malnutrition, Navdanya, New Delhi, 80pp

Shiva, V., and Singh, V. (2014), Wealth per Acre, Natraj Publishers, 230pp

Shiva, V. and Vinod, K. B. (2009), Climate Change at the Third Pole: The impact of Climate Instability on Himalayan Ecosystem and Himalayan Community, Navdanya/RFSTE, New Delhi, 223pp.

Shiva, V., Pandey, P. and Singh, J. (2004), Principles of Organic Farming, Navdanya, New Delhi Shiva, V. and Radha, H. B. (2001), Diversity The Hindustan way: An Ecological History of Food and Farming in India, Vol. 1 and 2, RFSTE/Navdanya.

Sinclair, M., Thompson, M., (2001), Cuba Going Against the Grain: Agricultural Crisis and Transformation. An Oxfam America Report, p. 5.

Singh, V. (2005), Agrobiodiversity, Sustainability and Food Security in the Himalayan Mountains: An Uttaranchal Perspective. Gorakhpur: Gorakhpur Environmental Action Group, p. 50.

Singh V., Shiva V. and Bhatt V. K. (2014), Agroecology: Principles and Operationalisation of Sustainable Mountain Agriculture, Navdanya, New Delhi, p. 64.

Srivastava, D., Swaroop, S., Mukerji, S., Roy, D. and Gautam, C.K. (2001b), Mass balance of Dunagiri Glacier, Chamoli District, Uttar Pradesh. Proceeding of Symposium on Snow, Ice and Glacier, March 1999, Abstract, 10-11.

Srivastava, D., Sangewar, C. V., Kaul, M. K. and Jamwal, K. S. (2001a), Mass balance of Ruling Glacier - A Trans-Himalayan Glacier, Indus Basin, Ladak. Proceeding of Symposium on Snow, Ice and Glacier, March 1999, Geological Survey of India, Special Publication. 53: 41-46.

Tangri, R. P. (2003), What Stress Costs. Halifax: Chrysalis Performance Strategies Inc.

The Ecologist, Vol. 13, No. 2/3, 1983. Tilman, David, 1999. *The Greening of the Green Revolution. Nature 396: 211-212.*

UNEP Climate Change Information Sheets, accessed on-line at: http://www.unep.org/dec/docs/info/ccinfokit/infokit-2001.pdf.

UNEP Report (2007), Fast Melting Glaciers Could Raise the Likelihood of Floods and Water Shortages, Uprety, D. C., "Rising nd Atmospheric Carbon Dioxide and Crops Indian Studies", 2 International Congress on Plant Physiology, Jan 8-12, 2003, New Delhi.

Vohra, C. P. (1981), *Himalayan Glaciers in the Himalaya: Aspects of Change*, (Eds. J. S. Lall and Moddie), Oxford University Press, New Delhi, pp. 138-151.

Wilson, R. J., Gutierrez, D., Gutierrez, J., Monserrat, V. J. (2007), 'An Elevational Shift in Butterfly Species Richness and Composition. Accompanying Recent Climate Change'. In Global Change Biology, 13: pp. 1873-1887.

WWF Climate Change. Nature at risk. Threatened species, accessed online at: http://www.panda.org/about_wwf/what_we_do/climate_change/problems/impacts/species/index.cfm.

WWF (2005), An overview of glaciers, glacier retreat, and subsequent impacts in Nepal, India and China. Available at http://assets.panda.org/downloads/himalayaglaciersreport2005.pdf.

WWF *Rivers at Risk report*, March 2007. http://assets.panda.org/downloads/worldstop10riversatriskfinal-march13.pdf.

Wright, J., (2005), Falta Petroleo! Cuba's experiences in the transformation to a more ecological agriculture and impact on food security. PhD thesis, Wageningen University, The Netherlands.

SYNERGETIC PRESS
SANTA FE ✦ LONDON